CAMBRIDGE COUNTY GEOGRAPHIES

General Editor: F. H. H. GUILLEMARD, M.A., M.D.

WESTMORLAND

Cambridge County Geographies

WESTMORLAND

by

J. E. MARR, Sc.D. F.R.S.

With Maps, Diagrams, and Illustrations

Cambridge:

at the University Press

1909

CAMBRIDGE UNIVERSITY PRESS
Cambridge, New York, Melbourne, Madrid, Cape Town,
Singapore, São Paulo, Delhi, Mexico City

Cambridge University Press
The Edinburgh Building, Cambridge CB2 8RU, UK

Published in the United States of America by Cambridge University Press, New York

www.cambridge.org
Information on this title: www.cambridge.org/9781107656758

First published 1909
First paperback edition 2013

A catalogue record for this publication is available from the British Library

ISBN 978-1-107-65675-8 Paperback

CONTENTS

ILLUSTRATIONS

The illustrations on pp. 5, 7, 9, 10, 13, 14, 27, 28, 33, 37, 39, 40, 45, 48, 57, 59, 81, 83, 87, 88, 92, 94, 101, 105, 106, 107, 108, 124, 126, 127, 129, 144 are from photographs taken by Mr W. B. Brunskill, of Bowness-on-Windermere, most of them specially for this book; those on pp. 32, 50, 55, 103, 104, 110, 112, 113, 115, 116, 118, 119, 136, 143 are from photographs taken by Mr Herbert Bell of Ambleside: those on pp. 12, 23, 24, 43, 52, 98, 111, 117, 120, 130, 137, 138, 141, 145, 146, 147 are reproduced from photographs in Frith's Series: the portrait on p. 139 is from a photograph by Mr E. Walker, and for the photograph from which the view on p. 58 is reproduced the author is indebted to Mr W. G. Fearnsides, M.A., Sidney Sussex College. The illustrations on pp. 101, 106 are from photographs of objects in the Kendal Museum taken by permission of the Museum Council.

1. Westmorland: Origin and meaning of the Word.

As Englishmen we are proud of our country, and we all know some of the reasons which led to the growth of the English nation and caused its people to occupy that particular tract of country which they to-day inhabit. Each of us, further, is proud of his native county. Many people of all ranks for example, young and old, take an interest in the annual struggle of the counties for supremacy in cricket. Yet comparatively few know the events which have caused our country to be separated into those divisions which we term counties. The irregular boundaries of these counties, which are so great a stumbling-block to the young student of geography, suggest that the causes which lead to the making of a county are by no means simple. At the present day, when divisions of a tract of land are made, they are often very simple. Look at the line which divides Canada from the United States. For a long stretch it is straight. Many of the smaller American divisions are bounded by straight lines. So in our country new towns like Barrow-in-Furness and Middlesborough are built with most of the streets in straight lines running at right angles to each other. In these cases the whole scheme of the parcelling out of the tracts is thought out before the division is made. But in the case of our counties there was no such principle of arrangement. They gradually grew up under varying conditions, and the boundaries were shifted more than once. These boundaries have usually been determined by some physical feature of the country which could be readily utilised, and often formed an actual barrier between adjacent divisions. As we shall see later, the county of Westmorland is separated from the adjoining counties along most parts of its borders by hill-ridges or by streams. Many divisions of the tract of country which we call Westmorland were made before its present boundaries were fixed. Indeed, it is best to avoid the word "fixed," for quite recently the boundary was altered in order to include the waters of the southern part of Windermere (formerly in Lancashire) within the area of our county.

All of us, even the most unobservant, must have remarked that the names of many counties end in "shire," as Lancashire, while others, as Westmorland and Cumberland, Kent and Essex, have not this ending. Shires are tracts of land which were divided by the Anglo-Saxons, the name itself being Anglo-Saxon, and meaning that the tract is due to the "shearing" or cutting up of larger tracts. Westmorland was never in its entirety a shire in Anglo-Saxon times, though a portion of it along and around the upper part of the Eden valley containing the town of Appleby was actually in those times regarded as a shire for a short period under the name of "Appelbischir." It was not however until early in the twelfth century that Henry I divided the old Cumbrian kingdom into two counties, namely Carleolúm and Westmarieland, the latter not differing very widely from the present Westmorland. The term county is from the old French word *comté*, a province governed by a count (*comes*), and it did not come into use till after the Norman Conquest. Such counties as Essex, Kent, and Sussex have kept their names, and roughly their boundaries as well, from the earliest times, and are survivals of former kingdoms.

The situation of Westmorland, which was for long on the borderland between Scotland and England, was the cause of the frequent changes in the boundaries of the political divisions which at different times existed in that region. The more important changes will be noted when we consider the history of the area, but in the meantime we must remember that the county in its present condition only came into being at a late date as compared with many other counties.

The name applied to the district by the Anglo-Saxons was originally Westmoringaland, "the land of the people of the western moors," in distinction from that of the people of the eastern moors, on the east side of the Pennine chain. The present name has not, however, been derived from that of Westmoringaland, but from Westmarieland or Westmerieland, used in the twelfth century, hence Westmerland. The meaning of this is land of the western *meres*, and not *moors*. Mere means a boundary as well as a lake, and it is doubtful whether the word as used here refers to lakes or boundaries. There is no doubt that

Westmerland is the more correct spelling; but Westmorland or Westmoreland has been so long in use that it must be retained. The principal writers have for a long time past adopted the spelling Westmorland, and to it we will adhere.

2. General Characteristics. Position and Natural Conditions.

Westmorland, situated far from the great centres of population, has played little part in the political and military affairs of the past, and the absence of coal and iron has prevented any part of it from growing into a great industrial district. It is essentially a pastoral county, and has been such for long ages. Sparsely populated, on account of the hilly nature of a large part of the county, the people have lived mainly in scattered homesteads or small hamlets in the valley-bottoms, taking their produce to the market towns and there obtaining the few necessaries for their simple lives which they could not produce in their own homes. The Westmorland people therefore are pre-eminently dalesmen.

Physically, the county is above all things a fell-tract, the lowland tracts, apart from the floors of the larger upland valleys, being confined to the southern portion, and to the upper part of the Eden valley. The fells do not as a rule rise into rocky peaks, but into round-topped ridges, usually occupied by extensive growths of peat or moorland grass, and it is only occasionally that the upper portions are formed of bare rock, this being chiefly the case with the limestone fells of the south.

Westmorland is a county of small swiftly-flowing clear streams, none of which are navigable. Part of it is included in the Lake District, but there are few large lakes, although the mountain tarns are fairly numerous. It cannot be described as a maritime county, although the

A round-topped hill : Hartsop near Ullswater

sea washes part of its southern boundary. At one time the port of Milnthorpe was of some importance, but its day as a port is now over.

There are no forests as the word is now understood ; there are forests, it is true, such as Ralfland Forest, west of Shap, but they are treeless. A forest is literally an open hunting ground, and that was the nature of the forests of Westmorland. Woods do occur, but they are not numerous, and most of them have been planted in recent times. But there is still much coppice in the lower parts of the valleys, with thick growths of hazel, birch, willow, alder, ash, and oak, and the coppices have had an important bearing upon the industries of the county in past times.

The climate is mild, and the rainfall is rather high, though, as we shall see in a later section, there has been much exaggeration concerning the amount of rain which falls in the Lake District.

The scenery of the county is varied, and much of it is very beautiful. The fell region of the north-western portion is especially fine, and the great scarp of the Pennine chain overlooking the lowlands of the Eden valley is impressive. From a picturesque point of view the palm must be given to the valleys. With few exceptions, the hills, owing to their rounded summits, are somewhat monotonous, but the valleys, with their admixture of crag, bracken-clad slope, and coppice, are very lovely. The waterfalls are miniature, but picturesque. The lakes present different types of beauty according to their surroundings, Ullswater being fine but rather sombre, while Windermere is almost park-like, owing to its open valley and the numerous houses with private grounds which fringe its shores and extend in places far up the valley sides.

Less familiar, but worthy of particular notice, is the scenery of the estuarine part of the river Kent. It is essentially of Lakeland type, and the sands add to the beauty of the scene.

There is much variety also in the river valleys. The

Stock Gill Force: Ambleside

open Vale of Eden, the gorge of the Lune below Tebay, and the various dales of the Lakeland portion, all have their characteristic features.

Little, then, as Westmorland has had to do with the development of the kingdom it is a county which well deserves our regard from its physical characteristics, and accordingly a book devoted to its description will not have much to say in illustration of political and commercial geography, but a great deal, on the other hand, concerning physical geography.

3. Size. Shape. Boundaries.

Westmorland, with Cumberland and the Furness district of Lancashire, form an upland region between the Pennine hills on the east, and the Irish Sea on the west, with the Solway and River Liddell on the north, and Morecambe Bay on the south.

The greatest length of the county is 42 miles, measured from a point on the sands of the Kent estuary south-west of Arnside to the river Tees nearly east of Cross Fell; and the smallest breadth along a line taken through the heart of the county is 18 miles, from the centre of Ullswater to a point in the Lune valley north of Lowgill Station. This narrow diameter is, however, exceptional, and is due to a tongue of Yorkshire projecting far to the north-west of the general boundary-line between that county and Westmorland, in the vicinity of Sedbergh. From west to east along a line drawn from Bow Fell at the

Ullswater

head of Great Langdale to the Yorkshire boundary east
of Stainmore the length is almost exactly 40 miles ; and
from north to south along a line from the Eden at Penrith
to the boundary with Lancashire just south of Burton-in-
Kendal is 33 miles. The county encloses an area of
505,330 acres, or 789 square miles.

**Hills forming County Boundary between Westmorland
and Cumberland**

There are only nine other English counties which are
smaller than Westmorland. It occupies about one sixty-
fourth of the entire area of England.

Comparing it with the counties which surround it we
find that Durham is half as large again as Westmorland,
Cumberland nearly twice its size, Lancashire considerably

over twice its size, and Yorkshire actually more than seven times as large.

The shape of Westmorland is very irregular, but if we leave the indentations out of account it forms a wide parallelogram with the angles at Bow Fell, the river Tees near Cross Fell, Black Hill in Stainmore Forest, and the point on the sands of the Kent estuary which we have already mentioned.

It will be advisable to follow the county limits carefully upon the map, and the variations in the heights should be noted, for the nature of this boundary is of great importance as bearing upon the history of Westmorland.

The northern boundary between the first two angles named does not run east and west but from 30° S. of W. to 30° N. of E. ; the next side is almost from N.N.W. to S.S.E.; the third is parallel to the first ; and the fourth nearly parallel to the second.

Along the northern boundary, Westmorland is throughout in contact with Cumberland. The line from Bow Fell runs along the watershed of the Lake District and on to a point of the Helvellyn range north of the summit of that name, where it leaves the high ridge, strikes eastward, and goes down Glencoin Beck to Ullswater, where that beck enters the lake some little distance from the lake head ; next it follows the middle line of the lake to the foot, and is carried along the river Eamont to its junction with the Eden. It passes up the last-named river for about three miles until Crowdundale Beck enters from the east, where its course extends to the head of this beck and across the

Brathay Bridge, joining Westmorland and Lancashire

Pennine watershed to Tees Head.   It then follows the course of the Tees to the junction of the three counties, Westmorland, Cumberland, and Durham.

Along the second line, the boundary follows the Tees for about eight miles, turns south-west for about two or three miles along Maize Beck, and hence extends south-eastward for about fifteen miles along an ill-defined course

Estuary of the Kent, looking seaward

on the east slope of the Pennine chain, crossing the various tributaries of the stream until the second angle is reached. This part of the boundary is purely artificial.

The third line at first follows the watershed of the Pennine range between the waters of the Eden and those of the Swale and Ure in Yorkshire as far as the head of Mallerstang; thence turns west and follows a tributary of

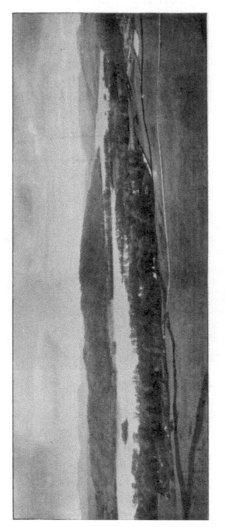

Windermere

the Rawthey, and that river itself to Cautley, where it leaves the stream, and extends along the watershed of the Howgill Fells, and drops down a tributary stream into the Lune between Tebay and Lowgill.   Here it turns south along the Lune, which it follows to the junction of the Rawthey and once more takes to the fell tops as far as Gragreth, where is the union of Westmorland, Yorkshire, and Lancashire.   Between the head of Mallerstang and this point is the large projecting tongue of Yorkshire. The boundary-line now turns nearly west, and is mainly artificial and ill-defined until the third angle is reached in the Kent estuary.

From the third angle the boundary leads north along the river Winster to the village of that name, and then west over the fell for two miles to Windermere. It now runs south along the east shore of that lake to its foot, and then along the whole of the western shore to the mouth of the river Brathay, up which it extends to Wrynose Pass, at the source of that river. From Gragreth to this point the boundary is between Westmorland and Lancashire.   From Wrynose it runs up the Lake District watershed for about four miles to Bow Fell, whence we started.

4.   Geology and Soil.

Before giving further account of the physical geography of the county it is necessary to learn somewhat of its geology, as the physical conditions are to a large extent dependent upon geological structure.

By Geology we mean the study of the rocks, and we must at the outset explain that the term *rock* is used by the geologist without any reference to the hardness or compactness of the material to which the name is applied; thus he speaks of loose sand as a rock equally with a hard substance like granite.

Rocks are of two kinds, (1) those laid down mostly under water, (2) those due to the action of heat.

The first kind may be compared to sheets of paper one over the other. These sheets are called *beds*, and such beds are usually formed of sand (often containing pebbles), mud or clay, and limestone, or mixtures of these materials. They are laid down as flat or nearly flat sheets, but may afterwards be tilted as the result of movement of the earth's crust, just as you may tilt sheets of paper, folding them into arches and troughs, by pressing them at either end. Again, we may find the tops of the folds so produced worn away as the result of the wearing action of rivers, glaciers, and sea-waves upon them, as you might cut off the tops of the folds of the paper with a pair of shears. This has happened with the ancient beds forming parts of the earth's crust, and we therefore often find them tilted, with the upper parts removed. Tilted beds are said to *dip*, the direction of dip being that in which the beds plunge *downwards*, thus the beds of an arch dip *away from* its crest, those of a trough *towards* its middle. The dip is at a low angle when the beds are nearly horizontal, and at a high angle when they approach the vertical position. The horizontal line at right angles to the direction of the dip is called the line of *strike*. Beds form strips at the surface, and the portion where they appear at the surface is called the *outcrop*. On a large scale the direction of outcrop generally corresponds with that of the strike. Beds may also be displaced along great cracks, so that one set of beds abuts against a different set at the sides of the crack, when the beds are said to be *faulted*.

The other kinds of rocks are known as igneous rocks, which have been melted under the action of heat and become solid on cooling. When in the molten state they have been poured out at the surface as the lava of volcanoes, or have been forced into other rocks and cooled in the cracks and other places of weakness. Much material is also thrown out of volcanoes as volcanic ash and dust, and is piled up on the sides of the volcano. Such ashy material may be arranged in beds, so that it partakes to some extent of the qualities of the two great rock groups.

The production of beds is of great importance to geologists, for by means of these beds we can classify the rocks according to age. If we take two sheets of paper, and lay one on the top of the other on a table, the upper one has been laid down after the other. Similarly with two beds, the upper is also the newer, and the newer will remain on the top after earth-movements, save in very exceptional cases which need not be regarded by us here, and for general purposes we may regard any bed or set of beds resting on any other in our own country as being the newer bed or set.

The movements which affect beds may occur at

different times. One set of beds may be laid down flat,
then thrown into folds by movement, the tops of the
beds worn off, and another set of beds laid down upon the
worn surface of the older beds, the edges of which will
abut against the oldest of the new set of flatly-deposited
beds, which latter may in turn undergo disturbance and
removal of their upper portions.

Again, after the formation of the beds many changes
may occur in them. They may become hardened, pebble-
beds being changed into conglomerates, sands into sandstones, muds and clays into mudstones and shales, soft
deposits of lime into limestone, and loose volcanic ashes
into exceedingly hard rocks. They may also become
cracked, and the cracks are often very regular, running in
two directions at right angles one to the other. Such
cracks are known as *joints*, and the joints are very important
in affecting the physical geography of a district. As
the result of great pressure applied sideways, the rocks
may be so changed that they can be split into thin slabs,
which usually, though not necessarily, split along planes
standing at high angles to the horizontal. Rocks affected
in this way are known as *slates*.

If we could flatten out all the beds of England, and
arrange them one over the other and bore a shaft through
them, we should see them on the sides of the shaft, the
newest appearing at the top and the oldest at the bottom.
Such a shaft would have a depth of between 50,000 and
100,000 feet. The beds are divided into three great
groups called Primary or Palaeozoic, Secondary or
Mesozoic, and Tertiary or Cainozoic, and at the base

of the Primary rocks are the oldest rocks of Britain, which form as it were the foundation stones on which the other rocks rest, and are termed the Precambrian rocks. The three great groups are divided into minor divisions known as systems.

In the following table (p. 20) a representation of the various great subdivisions or 'systems' of the beds which are found in the British Islands is shewn. The names of the great divisions are given on the left-hand side, in the centre the chief subdivisions of the rocks of each system are enumerated, and on the right-hand the general characters of the deposits of each system are recorded.

With these preliminary remarks we may now proceed to a brief account of the geology of the county, though to render it intelligible we must also say something about the geology of the adjoining tracts of Cumberland and of the Furness district of Lancashire.

In the county of Westmorland the following systems are found and are represented on the geological map at the end of the book: Recent and Pleistocene, Triassic, Permian, Carboniferous, Silurian, and Ordovician. The figure (p. 21) shews what is called a geological section drawn across the county from the foot of Ullswater to High Cup Nick and gives the arrangement of the rocks in the county. It represents what would be seen on the sides of a deep cutting if such were made through the county.

The actual Lake District consists chiefly of a mass of slaty rocks which occupy a tract of country about thirty miles in diameter of a roughly circular form. These rocks consist of three main groups, each composed of beds

	Names of Systems	Subdivisions	Characters of Rocks
TERTIARY	**Recent** **Pleistocene**	Metal Age Deposits Neolithic ,, Palaeolithic ,, Glacial ,,	Superficial Deposits
	Pliocene	Cromer Series Weybourne Crag Chillesford and Norwich Crags Red and Walton Crags Coralline Crag	Sands chiefly
	Miocene	Absent from Britain	
	Eocene	Fluviomarine Beds of Hampshire Bagshot Beds London Clay Oldhaven Beds, Woolwich and Reading Thanet Sands [Groups	Clays and Sands chiefly
SECONDARY	**Cretaceous**	Chalk Upper Greensand and Gault Lower Greensand Weald Clay Hastings Sands	Chalk at top Sandstones, Mud and Clays below
	Jurassic	Purbeck Beds Portland Beds Kimmeridge Clay Corallian Beds Oxford Clay and Kellaways Rock Cornbrash Forest Marble Great Oolite with Stonesfield Slate Inferior Oolite Lias—Upper, Middle, and Lower	Shales, Sandstones and Oolitic Limestones
	Triassic	Rhaetic Keuper Marls Keuper Sandstone Upper Bunter Sandstone Bunter Pebble Beds Lower Bunter Sandstone	Red Sandstones and Marls, Gypsum and Salt
PRIMARY	**Permian**	Magnesian Limestone and Sandstone Marl Slate Lower Permian Sandstone	Red Sandstones and Magnesian Limestone
	Carboniferous	Coal Measures Millstone Grit Mountain Limestone Basal Carboniferous Rocks	Sandstones, Shales and Coals at top Sandstones in middle Limestone and Shales below
	Devonian	Upper } Mid } Devonian and Old Red Sand- Lower } stone	Red Sandstones, Shales, Slates and Lime- stones
	Silurian	Ludlow Beds Wenlock Beds Llandovery Beds	Sandstones, Shales and Thin Limestones
	Ordovician	Caradoc Beds Llandeilo Beds Arenig Beds	Shales, Slates, Sandstones and Thin Limestones
	Cambrian	Tremadoc Slates Lingula Flags Menevian Beds Harlech Grits and Llanberis Slates	Slates and Sandstones
	Pre-Cambrian	No definite classification yet made	Sandstones, Slates and Volcanic Rocks

Geological Section from Ullswater to the Pennine Hills

5. Trias. ⎫
   ⎬ New Red Sandstone.
4. Permian. ⎭

3. Carboniferous.

2. Volcanic Series ⎫ Ordovician.
1. Skiddaw Slates ⎭

x. Whin Sill.

F F. Faults.

many thousands of feet in thickness. The oldest group, called the Skiddaw slates, is formed chiefly of clay slates with a few sandstone Sands. The middle group, the Borrowdale Volcanic Series, is made up of lavas and ashes of various characters. The upper group is very variable. Like the lower it consists of sediments, with impure limestones at the bottom followed by clay slates and sandstones. The rocks are very old and are known as the Ordovician and Silurian rocks, which are amongst the oldest of the British Isles, or indeed of the world. The upper and lower groups contain many fossils. These fossils are interesting for several reasons, among others, because they give us some idea of the forms of life which existed in these early times—forms differing in many ways from those of the present day. Interesting as these fossils are, however, they are of slight importance as regards the geography of Westmorland, and we may pass them by with the bare notice of their occurrence.

These old slates of the Lake District after their formation were thrown into a great arch with the centre of the arch passing through the Skiddaw group of hills and the southern part of the arch sloping southward, so that the beds sink down into the ground as a whole in a southerly direction. Accordingly the oldest group of slates is found chiefly in the northern part of the district, the middle group in the central tract, and the youngest group in the southern portion.

During the time of the formation of the great arch the topmost rocks of the arch were removed by rivers, and probably by the waves of the sea, and a comparatively flat tract was formed by these processes of planing down. At the same time the rocks were much compressed and hardened, and the finer ones turned into slates. Also igneous

**Sour Milk Gill, near Grasmere,
shewing tract formed of slate-rocks**

rocks were squeezed into them and cooled, of which the most important mass in Westmorland is the well-known granite of Shap Fells.

On the levelled surface of these old rocks another set

of beds was formed, belonging to what is known as the Carboniferous system, so called because it contains productive coal. Unfortunately for Westmorland only the lower part of these beds are found in the county with one exception, the upper parts, in which the valuable coal-seams occur, being absent save at a place east of Brough-under-Stainmore, where a very small patch occurs.

Carboniferous Limestone, Scout Scar, near Kendal

The most interesting of the lower beds are the masses of white limestone, which from its frequent existence in hills is spoken of as the Mountain Limestone. This contains a great number of fossils in places, and is indeed largely composed of them.

After the formation of the Carboniferous rocks, another movement took place, which we need not notice fully, as it was less important than the earlier one. At this period was probably injected a mass of igneous rock known as the "Whin Sill," lying nearly level among the limestones of the Pennine hills.

Another set of rocks consisting chiefly of red sandstone was laid upon the older rocks. These belong to the New Red Sandstone age, so called to distinguish it from that of the Old Red Sandstone, which is practically unrepresented in Westmorland.

Long after this the Lake District was affected by another uplift, which was not saddle-shaped but dome-shaped, and as the upper rocks were again swept away we find the old slate rocks now on the surface in the centre of the dome with the Mountain Limestone forming an almost complete ring around them, and outside this ring of limestone is another, less perfect, of New Red Sandstone. On the east side of the New Red Sandstone of the Eden valley the Carboniferous rocks again appear at the surface in the Pennine hills, for the Eden valley between the Lake District and the Pennine hills is a geological trough.

With these descriptions and a study of the geological map we can understand the structure of Westmorland, which forms the south-eastern portion of the dome.

The western part of the county is formed of the old slate rocks, which may be traced along the county boundary from the estuary of the Winster in the south to the foot of Ullswater on the north. Here we meet with the junction with the ring of Carboniferous beds which form the strip of country on the west side of the Vale of Eden as far to the south-east as the fells south of Ravenstonedale, where the ring passes into Yorkshire; hence the county boundary from this southward is through Silurian rocks as far as the district near Kirkby Lonsdale, where it again cuts into the ring of Carboniferous rocks, and continues therein to the Winster estuary. The northern outline of the ring in the south of Westmorland is very irregular, and a great tongue of Mountain Limestone runs northward through the Silurian rocks to Kendal, and on its western side crosses the estuary of the Kent to form Whitbarrow.

The New Red Sandstone occurs in Westmorland in the upper part of the Vale of Eden, forming a narrow triangle with its apex near Kirkby Stephen, and east of this, as already stated, the Carboniferous rocks again come to the surface, and form the Westmorland portion of the Pennine hills.

Of that part occupied by the slaty rocks, the northern part of the slate tract as far south as a line drawn from Shap Wells to the head of Windermere is made of the rocks of the Volcanic group, save a little strip parallel to the Carboniferous rocks between the foot of Ullswater and Shap which (owing to complications of structure which we need not consider) is formed of the Skiddaw slates. The rest of the slate tract south of this is composed of the upper group of slates.

Since the last uplift forming the dome, the work of rivers and glaciers has largely been concerned in cutting out the valleys, leaving the intervening portions to project as the fells. Much of the work has been done by the rivers, which are able to saw their way downwards, thus deepening the valleys, while rain, frost, and the other agents of the weather cause the material of the valley-sides to be carried downwards to the streams at the base, thus widening the valleys. At a time which as compared with the formation of the rocks we have described is

An Eroded Valley, Troutbeck, Windermere

but as yesterday, though very remote as compared with the beginning of the human history of our land, the district was occupied by masses of ice moving downwards from the upland regions towards the sea, and these masses of moving land-ice left their well-known characteristic marks on the rocks, which are rounded and polished by their action. A celebrated example of this, figured by the great geologist, Sir Charles Lyell, is seen near Ambleside church. In addition to this, the ice helped to increase the depth and width of the valleys, and much of the material which it ground down and carried away it left in sheltered spots and lowland tracts to form the stiff clay, sometimes mixed with sand, and containing blocks of

Glaciated Rock, Bowness

stone of various sizes, which is known as boulder clay The lakes of the district occur in hollows partly due to excavation by this ice and partly to blocking of the valleys by deposits of boulder clay or similar material, some lakes being due entirely to one process, some to the other, and others again to a combination of the two

Since the glacial period, the action of the weather has

caused the upper surfaces of the rocks to be broken up into pieces of various sizes, and parts of the glacial accumulations to be loosened, giving rise to soils. Of these there are four main types, dependent for their characters to a large degree upon the nature of the underlying rocks. In the slate tracts, character of the soil is largely dependent upon the glacial accumulations, which have in so many places covered the slaty rocks; where the latter are uncovered by glacial materials, they are often bare of soil. The glacial materials give rise on the whole to a poor stiff stony soil, usually wet, though where much sand occurs in the glacial masses, the soil is looser and drier.

The Mountain Limestone when not covered by glacial materials is usually bare ; here and there a short sweet turf occurs. Where glacial accumulations lie thickly over the limestone, the soil naturally resembles that of the slate tracts, but where the glacial materials are thin, a fairly rich soil may be produced.

The third type of soil is formed over the New Red Sandstones of the Eden valley. This is often a light sandy loam of a red colour, but on this tract also variations are produced by the presence of glacial materials.

The fourth type is found occupying the sites of former lakes which have been filled in by gravel and silt, and also in the upper part of the estuarine flats of the river Kent. The old lake-sites are scattered over the county. On these flat tracts there has been as a general rule an abundant growth of peat, which yields a rich black soil. With the peat is mixed a variable amount of silt, which causes the soil to be especially valuable.

5. Surface and General Features.

Westmorland may be divided according to its physical structure into three important divisions and three minor tracts.

Of the important divisions the largest is the slate tract, which occupies more than half the county, and lies to the south-west of a line which may be drawn in a general north-west and south-east direction from the foot of Ullswater to the fells south of Ravenstonedale.

The second division forms the Westmorland portion of the Vale of Eden, in the shape of a rough triangle with its base along the county boundary between Penrith and a stream south of Cross Fell, and its apex near Kirkby Stephen.

The third division consists of that part of the Pennine chain which lies within the county. This tract is to the east of that just described, and extends to the county boundaries dividing Westmorland from Durham and Yorkshire.

Of the less important tracts, the first separates the Eden valley from the slate tract, and lies north-east of the line separating that tract from the area of newer rocks. Near Ravenstonedale it bends round to the east to join the Pennine hills around the head waters of the Eden. The second tract is marked by the Mountain Limestone rocks of the fells about Kendal, Whitbarrow, and Burton, and we include in it the country around Kirkby Lonsdale. The third consists of the low ground around the estuary of the Kent.

We will now shortly consider the general characters of these divisions, beginning with those of most importance.

The *first* division, the slate tract, is essentially a region of fells—high ridges separating the various valleys which have been carved out from the upland tract.

The fells as a whole are highest in the northern part of the slate tract and gradually become lower as we pass southwards, though there are exceptions.

The highest fell tract of the county forms a horse-shoe-shaped mass at the head of Ullswater, with the open portion of the shoe facing the lake. The extreme north of this horseshoe-shaped elevated tract lies in Cumberland, but it enters Westmorland at Stybarrow Dodd, to the north of Helvellyn, and for some distance to the south the county boundary runs along its summit. For many miles the tract rises to a height of over two thousand feet, save where cut through by passes, and the following important hills occur along it, as one goes from the north-western projection to that on the north-east: Helvellyn (3118 ft.), Fairfield (2863 ft.), Red Screes (2541 ft.), High Street (2663 ft.), High Raise (2634 ft.). From High Street a spur extends eastward to Shap Fells, on which is situated Harter Fell (2539 ft.).

Another tract of high ground extends along the county boundary in a general south-westerly direction from Dunmail Raise Pass to the Three Shire Stones on Wrynose Pass. The highest point on this line of fells is Bow Fell (2960 ft.), which with the Crinkle Crags makes the fine screen of hills at the head of Great Langdale.

Red Tarn and Helvellyn

Connected with this tract and south of the county boundary are the beautiful Langdale Pikes (2401 ft.).

The above high tracts are entirely composed of the hard resisting rocks of the Borrowdale volcanic group.

The Upper Slate group is on the whole marked by

Langdale Pikes: a wintry scene

lower fells, but on the east side of the Lune the fells formed of the rocks of this group rise to a height of 2220 ft. in the hill called the Calf on the county boundary between Westmorland and Yorkshire.

The *second* division is important not so much on account of its area (for if we include only the portion situated on the New Red Sandstone it is exceeded in size by one of the less important tracts), but because of its physical characters, which have played a great part in determining the settlement of man in the county. It is essentially low ground, between the Lakeland heights on the west, and those of the Pennine range to the east and south. The greater part of it lies at a height of less than 500 feet, and in comparison with the uplands by which it is surrounded it constitutes an undulating plain. Though usually spoken of as part of the Eden valley, it is not so in the strict sense, the actual Eden valley being a narrow strip on either side of the banks of the river Eden towards the centre of this undulating plain.

The *third* important division—the Westmorland part of the Pennine hills—is composed of Carboniferous beds which sink gently downwards towards the east. These rocks are separated from the New Red Sandstone rocks as far south as Kirkby Stephen by a great earth-fracture which comes to the surface at the base of the west side of the Pennine hills, and the resistant rocks of those hills stand out with a steep scarp facing westward, so that the summit ridge of the Westmorland part of the hills usually lies only about two miles east of the base. From this ridge, the Pennine hills slope gently eastward into the counties of Durham and Yorkshire. South of Kirkby Stephen the narrow upper Eden valley is carved entirely out of Carboniferous rocks, and lies in the range and not on its western side.

In part of the Pennine range, along the line between the northern county boundary and Roman Fell, south-east of Appleby, a thin strip of slate rocks lies between the New Red Sandstone of the Eden valley and the Carboniferous rocks of the Pennines. It gives rise to a series of conical hills in front of the great Pennine scarp. The latter rises into very high ground, usually over two thousand feet above sea-level, save where indented by passes. The highest point of this tract is the western end of the summit of Mickle Fell, north of Brough-under-Stainmore, which on the county boundary attains 2547 ft. above sea-level. The actual top of the fell—which is in Yorkshire and is indeed the highest point in that county —is 2591 ft. high.

Turning now to the three less important tracts, the part composed of Carboniferous rocks lying between the slate district and the Eden valley may be regarded as a miniature Pennine chain, with a steep scarp facing westward and gentle slopes sinking eastward to the Eden valley.

The roughly triangular patch of Mountain Limestone extending from near Kendal to the southern boundary of the county is mainly high ground, consisting of fells chiefly of bare rock, usually having steep cliffs facing to the west and gentler slopes to the east. The principal fells are Kendal Fell and Farleton Knott on the east side of the low ground of the Kent estuary, and Whitbarrow to the west side. No part of this tract rises to a height of 1000 feet, except in the district around Kirkby Lonsdale, where high fells occur on the east side of the Lune.

Lastly we have the flat tract of peat moors, reclaimed land, and sand-banks of the estuary of the Kent, extending southward from near Underbarrow to the county boundary with Lancashire, which crosses the sands south of the

Furness Railway viaduct near Arnside. Save for a few rocky masses projecting through it, this tract is a flat which is little above high tide, the sand-banks indeed being below.

6.   Watersheds and Passes.

Westmorland having so much high land, the watersheds have had a most important effect as barriers checking the spread of the people who sought to enter the area from elsewhere at various times, while on the other hand the passes have allowed them to pierce these barriers, and gain access to lower ground on their further sides. It is necessary therefore that we should consider these barriers and their passes in some detail, and that they should be studied on the map.

There are two important watersheds in Westmorland which are connected with one another so as to form a rough letter T laid on its side thus ⊣. Of these the part of the ⊣ on the east is the Pennine chain, which here forms the watershed dividing the rivers flowing eastward into the North Sea from those flowing westward into the Irish Sea. The other part separates the rivers of the Lake District which flow northward and north-westward into the Solway and that part of the Irish Sea north of St Bees Head from those which flow southward into Morecambe Bay.

The low grounds of Westmorland, which on account of their comparative fertility were the tracts desirable to the early invaders of the territory, lie in the angles where

Grasmere and Dunmail Raise

the two lines forming the ⊣ join. The northern angle is occupied by the Westmorland lowlands of the Eden valley, which extend also over the Carboniferous rocks south of the Eamont valley to the foot of Ullswater, while the southern angle contains the lowland tract between Kendal Burton-in-Kendal and Arnside, continuous with that of the Lune valley above Kirkby Lonsdale.

The low grounds on the north are continuous with those of the Cumberland plain, which could be fairly easily reached by three routes (1) from the sea; (2) from the south of Scotland; (3) through a pass between Carlisle and Haltwistle, between the Tyne and Eden valleys.

Similarly the lowlands of southern Westmorland could be reached (1) from the sea; (2) from the Lancashire plains to the south; (3) through a pass across the Pennines between Skipton and Settle in Yorkshire.

Apart from the two passes above named, and one to be noticed more particularly below, the Pennines form a continuous tract of high ground which in early days served as a most important barrier between the dwellers on the east and west sides of the north of England.

Let us now turn to that part of the Pennine chain which is situated in Westmorland. We have seen that on the east side of the Eden valley the county boundary lies some way down the eastern slopes of the chain, so that the actual watershed is in Westmorland. At the head of the Eden valley the boundary runs across high fell ground in the heart of the Pennines, and where it passes from these hills to the Lune and its feeders the

Pennine hills in that part of Yorkshire lying due east still form an effective barrier.

In one place however the Pennine barrier is broken by a most important pass—Stainmore, the summit of which lies in Westmorland a short distance west of the boundary between the county and Yorkshire. Although the fells

Dunmail Raise, summit

on either side within a very short distance rise to heights of over 2000 feet, the pass is only about 1400 feet above sea-level, and as it connects the Eden valley with that of the Tees, and hence forms an easy link between the low grounds of the Eden and those of north-eastern England, it has again and again been used as a highway.

To the south of Stainmore is a somewhat higher pass

between the Eden valley and Swaledale which never attained the importance of Stainmore. Still further south is another pass between the Eden valley and Wensleydale, and as still another pass a little further south separates Wensleydale and Garsdale, the latter occupied by a river draining into the Lune, this tract of relatively low ground

Shap Pass (marked by a x) from the South

is important, having recently been used for the route of the Midland Railway.

Let us now turn to the watershed which separates the northern from the southern part of the Lake District. Starting at the western portion of this watershed where it enters Westmorland, at the Three Shire Stones, we find that it is there situated on a pass (Wrynose, 1270 ft.)

which separates the Coniston group of fells in Lancashire from the Bow Fell group. Wrynose Pass is one of much importance, as we shall see later. About four miles north of Wrynose is Rossett Gill Pass (2106 ft.), and less than two miles east of this is another, the Stake (1581 ft.), both of importance to the tourist, but so high that they were probably quite unimportant in early days. About five miles north-east of the Stake is a pass of very great importance, namely Dunmail Raise. It is only 783 feet above sea-level and has at all times formed a route between the north and south of the Lake District. Between Wrynose and Dunmail Raise the watershed has separated Cumberland from Westmorland, but the county boundary here turns north along the Helvellyn range, and soon, as already stated, leaves the fells to follow the streams down to Ullswater. The watershed runs eastward, and accordingly the passes now to be named are situated in Westmorland and form connexions between the lower grounds on the north and south of the county. The first, Grisedale Pass (1929 ft.), connects the Grasmere and Ullswater valleys. About four miles south-east lies a much more important pass, Kirkstone (1481 ft.), connecting the Windermere and Ullswater valleys. Four miles east of Kirkstone is the Nan Bield (2100 ft.), between the Kentmere and Haweswater valleys, and less than two miles eastward again the Gatescarth Pass (1950 ft.), between Long Sleddale and Haweswater. Walking nearly eastward from the Nan Bield we reach the most important pass of the county, Shap Summit, just under 1000 feet above sea-level, and connecting the lowlands of the Eden valley on the north with the Lune valley on the south. A pass somewhat east of this connects Tebay and Appleby, but all the tract of ground between this and Shap Summit may be regarded as the Shap Pass. Two other passes connect the Lune and Eden drainage areas. One over Ash Fell, between Ravenstonedale and Kirkby Stephen, is in reality a double pass, for the railway, instead of going over the fell, is carried through a most remarkable narrow gorge near Smardale Station. The other by Rawthey Bridge connects Sedbergh with Kirkby Stephen.

There are, of course, many minor watersheds separating vale from vale, with passes over them, but they have played little part in the distribution of the population and require no notice.

7. Rivers and Lakes.

Five of the rivers of Westmorland flo v directly into the sea, though only one of these reaches the sea within the confines of the county : all the other riv s are tributary streams. We have noted that the head-waters of the Tees form the county boundary for about eight miles, between Tees Head and Cauldron Snout, and along this part of the river's course it receives a few tributaries from that part of the Pennine watershed which there lies in Westmorland. Accordingly several square miles of this distinctly western county drain into the North Sea.

The chief rivers on the western side of the Pennines are the Eden on the north side of the Lake District watershed, and the Lune on the south side. The other rivers reaching the sea are the Leven and the Kent. The latter is the only main river which has its course entirely within the county, and the Leven itself is not in the county, though some of its feeders are. The Eden rises in the Pennine fells some ten miles south of Kirkby Stephen, and runs at first in a narrow valley—Mallerstang—flanked

The Lune

by high hills, Mallerstang Edge on the east and Wildboar Fell on the west. Shortly before reaching Kirkby Stephen the river flows from the Carboniferous rocks on to those of New Red Sandstone age, and when reaching these softer rocks, the valley suddenly widens and the plain of the Eden is entered. The river continues in this wide tract, though the actual river valley is sometimes narrow, as at Appleby, which is situated on a loop of high ground nearly surrounded by the river. About 30 miles from its source the river enters Cumberland.

The tributaries of the Eden are mostly short. Those coming from the east descend from the Pennine heights. The most important are the Belah with its tributary Argill Beck from Stainmore, Hilton Beck and High Cup Gill from Hilton and Murton Fells, Knock Ore Gill from Milburn Fell, and Crowdundale Beck along which the Cumberland county boundary passes, coming from Cross Fell.

From the west side, Scandale Beck comes out of Ravenstonedale and joins the Eden north of Kirkby Stephen, and the Lyvennet rises in the fells south-east of Shap and runs northward for about fifteen miles to the Eden. The principal tributary however is the Eamont, which joins the Eden about four miles east of Penrith, and forms the county boundary from its point of issue from Ullswater to its junction with the Eden.

The waters of the Eamont rise in the numerous valleys of the Helvellyn and High Street ranges, and flow into Ullswater. From Kirkstone Pass to the junction with the Eden is a distance of about 30 miles. Before reaching the Eden, the Eamont receives from the south the river Lowther, coming from Wet Sleddale near Shap, and having several tributaries, including Haweswater Beck which flows from the lake of that name.

The river Lune rises in Ravenstonedale, a horseshoe-shaped hollow in the hills. Several streams flow south-wards from the Howgill Fells into this hollow, of which

the more easterly run through the gorge at Smardale into
the Eden, while those further west unite to form the
Lune, which flows westward for a few miles through a
fairly wide valley to Tebay. Here it receives a feeder,
the Birkbeck, from Shap Fells, and turns due south
through a gorge among high hills to Sedbergh. A little

The Kent, below Kendal

distance south of Tebay it receives a tributary from the
north-west from the Borrowdale valley, and shortly afterwards forms the county boundary with Yorkshire until the
Rawthey, a Yorkshire stream, enters from the east, after
which the course of the Lune is again wholly in Westmorland in a fairly wide valley to a point just south
of Kirkby Lonsdale, where it enters Lancashire. The

course of the Lune, where the river flows partly or entirely through Westmorland territory, has a length of about 30 miles.

The Kent rises in Kentmere 12 miles north of Kendal among the hills of the High Street group. It is curious to note that the valley here has taken the name of the mere or lake, now no longer in existence, while the principal town is known by the name of the valley. These names are due to the tendency of the inhabitants to shorten words. Kentmere stands for Kentmeredale, while Kendal is in full Kirkby-in-Kendal (or Kent dale). The river reaches the estuary about six miles south-south-west of Kendal, and the estuary itself extends over another six miles to its termination in Morecambe Bay.

Of the tributaries of the Kent (including the estuary as part of the river) the highest stream of any importance —the Sprint—occupies the upland valley of Long Sleddale. This valley, which resembles Kentmere, lies to the east of Kentmere and parallel to it. The Sprint joins the Kent over a mile north of Kendal, and a few hundred yards lower another tributary, the Mint, which flows from Bannisdale, an upland valley parallel to and east of Long Sleddale, also enters the Kent. At the head of the estuary a small stream, the Gilpin, joins the Kent, and a little lower the Bela, a stream not many miles long, enters from the east at the old port of Milnthorpe. Lastly, the Winster, which is separated from the Kent estuary by Whitbarrow Scar, joins that estuary between Grange-over-Sands and Arnside. This river for several miles marks the county boundary.

The Leven is situated entirely in Lancashire, and has a very brief course, flowing tumultuously down a narrow valley between the foot of Windermere and the head of its estuary. The feeders of the Leven, however, are of great importance. The most distant tributaries come from the southern sides of the Bow Fell mass of hills, and join to form a stream which flows down Great Langdale. At Elterwater this is joined by the Brathay coming from Wrynose Pass, through Little Langdale. The waters of the Brathay are shortly joined by those of the Rothay, which drains the Grasmere valley and receives numerous tributaries from the Fairfield fells. Immediately after receiving the Rothay waters, the Brathay enters the head of Windermere. Along the whole of its course the Brathay forms the county boundary between Westmorland and Lancashire.

One fairly important stream enters Windermere on its eastern side, namely that which occupies the Troutbeck valley. This starts among the High Street Hills, and flows southward to the lake, which it enters two and a half miles from its head.

Westmorland possesses many lakes and tarns. The term tarn as a whole is applied to a small lake, usually less than half-a-mile in length. The larger lakes occupy parts of the floors of the valleys, while the greater number of the tarns are perched in hollow combes on the hill sides far above the valley bottoms, and in many cases the streams which come from them flow in cascades down the sides of the larger valleys.

One large lake only has the whole of its waters

Rydal Water

included in Westmorland, namely Windermere. This is the largest lake not only in the Lake District but in England. Ullswater, another of the large lakes of Lakeland, has its upper reach entirely in Westmorland, but the county boundary between Cumberland and Westmorland runs along the centre of the two lower reaches, so that their south-eastern shores only are in Westmorland. Other lakes of the county arranged according to size are Haweswater, Grasmere, Rydal, Elterwater, and Brother's Water. The two latter are rather tarns than lakes but are named here because they occur along the larger valley floors.

A considerable number of tarns are found in the uplands. We will mention the principal, arranged according to the valleys to which they belong. Towards the head of the Ullswater valley we have Keppelcove, Red, and Grisedale tarns in combes of the Helvellyn fells, and Hayeswater and Angle tarns among the High Street fells. At the head of the Haweswater valley are Bleawater and Smallwater in the crags of High Street. Kentmere possesses the dreary Skeggleswater. Codale and Easedale tarns belong to the Grasmere valley, and Stickle, Blea, and Little Langdale tarns to the twin Langdale valleys.

Let us now consider the size and depths of the lakes. Windermere has a length of 10½ miles, and covers an area of $5\frac{7}{10}$ square miles. Its maximum breadth is just under a mile opposite Millerground Bay, and the average breadth just over half-a-mile. It is 130 feet above sea-level, and drains an area of nearly 90 square miles. The greatest

Haweswater

depth is 219 feet, at a distance of a mile and a half from the lake-head. The lake runs nearly north and south, but the upper part bends slightly westward towards the head. The shores are not greatly indented, Pullwyke, near the head on the eastern side, being the most marked bay: others are the bay near the Ferry Hotel on the west side, and Lowwood, Millerground, Rayrigg, and Bowness bays, and those at the Ferry Nab and Storrs on the east side. There are two fairly large deltas, namely that of Trout-beck on the east side and of Cunsey Beck on the west. Many islands or "holms" occur; a few small ones near the head and towards the foot, and a cluster of large and small together opposite Bowness Bay, the largest being Belle Isle, nearly three-quarters of a mile in length, but narrow.

Ullswater is about $7\frac{1}{3}$ miles long measured along the central line of the lake, and has an area of nearly $3\frac{1}{2}$ square miles, its greatest breadth is 1100 yards and the average breadth over 800 yards. It is 476 feet above sea-level and drains an area of 56 square miles. The greatest depth is 205 feet, at a point about $1\frac{3}{4}$ miles from the head. The lake extends on the whole from south-west to north-east but is far from straight, having a rude Z-shaped form.. The upper or southern reach is the shortest, being about one mile long, and stretching from south to north. The other reaches are each about three miles in length, the middle reach lying about west-south-west—east-north-east, and the northern more nearly south-west—north-east. There is one important bay, Howtown Bay, at the head of the northern reach. The

Blea Tarn, and Langdale Pikes

Glenridding Beck has built a large delta near the lake-head, and Sandwick Beck another on the southern side of the middle reach. A few small rocky islets occur in the upper reach.

Haweswater is only about $2\frac{1}{3}$ miles long, with an area of a little over half a square mile ; the average breadth being slightly over 400 yards, and the greatest breadth 600 yards. It lies at a height of 694 feet above the sea and is the highest of the valley lakes of the Lake District. It drains an area of over 11 square miles situated entirely in Westmorland. The greatest depth, 103 feet, is near the middle of the upper reach. The lake lies nearly south-west—north-east, and has two reaches, separated by the delta of Measand Beck, which nearly divides the lake into two somewhat below its middle. The south-eastern shore is singularly straight, that on the north-west more indented.

The other valley lakes are much smaller than those noticed. Grasmere is about one mile long and about half-a-mile across at the broadest part, with a depth of 180 feet. Rydal is even smaller and much shallower, being only 55 feet deep, while Elterwater and Brother's Water are yet smaller.

Of the upland tarns of the county the largest is Hayeswater with a length of half-a-mile, the highest is Red Tarn on Helvellyn (2356 ft.), and perhaps the most beautiful and characteristic is Bleawater, on the eastern side of High Street.

8. Scenery.

In a county of which the scenery is an important factor affecting the inhabitants, some attention must be paid to the causes and character of that scenery. These are dependent partly upon the geological structure, partly upon meteorological conditions whether acting directly— (e.g. the effects of sunlight and clouds upon the view)— or indirectly—(as affecting the vegetation, and also the agents such as frost, rivers, and glaciers, by which the details of the scenery have been largely determined).

At the outset we may take into account the effects of the more important rock-groups in controlling the nature of the surface.

Beginning with the slate-rocks, the oldest group, the Skiddaw slates, is of little importance in this county, though its importance is very great in Cumberland. The Skiddaw slates of Westmorland are the softest beds of the group, and are readily worn down to a stiff clayey wet soil supporting a marshy type of vegetation, hence the ground occupied by these rocks in the Lowther valley and between this and the foot of Ullswater is an uninteresting marshy tract of a flattish character.

The rocks of the middle division of the slates, belonging to the Borrowdale volcanic group, undoubtedly afford the wildest scenery in the county. They are the hardest of the rocks which are extensively found, and as there is considerable variety in their hardness, the lavas and some of the volcanic ashes being peculiarly hard, while some of

the ashes are softer, a considerable diversity of outline is
thus caused by these alternations. Again, they are affected
by very regular systems of gigantic cracks or joints, often
with belts of smashed rock along the cracks, and these
cracks have been lines of weakness which have frequently
been worn into notches and gorges, while they also define
the sides of cliffs and of rock pinnacles. To the hardness
of the rocks we owe in a great degree the superior eleva-

Langdale Pikes

tion of the fells which are composed of them ; and to the
variations of hardness and the nature of the joints the
frequent alternation of cliff and slope which is so marked
a feature of these fells, and is displayed at its best in the
Langdale Pikes, though well shewn in most of the hills
seen in the views from around the heads of Windermere,
Ullswater, and Haweswater. On some of the more level
ridges this type of scenery is replaced by smoother outlines,

due to the abundant growth of peat, as for instance on the fells east of Langdale Pikes visible from Windermere.

The upper slates give rise to tamer scenery, for the rocks are not so hard as those of the Borrowdale group, and there is less variety in the degree of hardness. Accordingly they rise on the whole to less elevation, and though crags are numerous they are on a smaller scale than those of the rocks just noticed, while the peat-covered uplands are more frequent. This type of scenery is well displayed along the sides of Windermere and in the country to the east and south-east of that lake. The highest hills formed of these rocks are the Howgill Fells, where however the peat growth is very marked and bare rock less frequent than in other tracts occupied by the upper slates.

The Carboniferous rocks are chiefly limestone (though above the limestone we find in some places Millstone Grit), and in this group is the mass of "whin sill" noticed in the geological section, which is responsible for some interesting scenery in High Cup Gill and the upper part of the Tees valley.

The structure which is presented by the Mountain Limestone hills has been well named "writing-desk structure," for the gently inclined beds form gentle slopes, with steep cliffs determined by the nearly vertical joints on the other sides of the hills. This structure is typically shewn in Farleton Knott, Arnside Knott, and Whitbarrow, and on a large scale in the Pennine chain east of the Eden valley, and the strip of Mountain Limestone between the Lowther and the Eden valley. The bare white cliffs and fissured "clints" offer a marked contrast to the surfaces of the fells formed out of the slate-rocks.

The clints just mentioned are flat or fairly flat surfaces of limestone with fissures produced by the widening of the vertical joints by acidulated rain-water. This water is capable of dissolving the limestone, and the bare flat or gently-sloping limestone surfaces are therefore often

Whitbarrow: a limestone hill shewing "desk-structure"

traversed by two sets of fissures at right angles to each other, penetrating for many feet or even yards from the surface. The sides of these are often honeycombed by the solvent action of the rain assisted by the vegetation that may grow abundantly within.

The massive well-jointed Millstone Grit, when found on fell-tops produces flat-topped hills with steep scarps or cliffs beneath, well shewn in Mallerstang Edge and Wildboar Fell on either side of the Eden valley.

The rocks of New Red Sandstone age, being soft, are readily worn away, hence the comparatively low ground occupied by them in the Eden valley. Furthermore, their rapid breaking-up allows abundant formation of soil, and the bare rock is seldom naturally exposed save along the

High Cup Nick, shewing cliff due to "whin sill"

sides of some gorges. In addition to this a great deal of glacial material has accumulated in the low ground of the Eden valley and has masked the rocks beneath.

The effect of the ice of past ages in hollowing the valleys has been noticed in the geological section. We are here concerned with its deposits. We may first notice the little moraines which were left by the upland glaciers

in the more central valleys as piles of rubbish at their
ends. They consist of hummocky hillocks of clay,
gravel and stones, covered with coarse vegetation, and

Easedale Tarn, shewing glacial moraine beyond
Tarn and below hill

produce a somewhat desolate effect. These moraines may
be well seen in several of the upper valleys whose waters
drain into Ullswater; in Fordingdale Beck draining into
Haweswater; in Swindale; and in a number of the valleys

which drain into Windermere, as for example Woundale near Kirkstone Pass, the Easedale Tarn valley, and Mickleden at the foot of the Rossett Gill Pass. The last two examples are especially good. A large number of the tarns are partly held up by these moraines.

The boulder-clay which was noticed in the geological section is spread widely over the low-lying grounds. It is often arranged so that its upper surface forms parallel mounds like the backs of whales. These are known as "drumlins." They occur abundantly in the low ground between Staveley and the south of the county, and also in the Eden valley. Kendal Castle stands on one of them.

Latest of the deposits which have produced large effects upon the scenery are those which fill in lakes and the estuarine tracts at the mouth of the Kent. Long flats occur at the heads of the principal lakes where the rivers are pushing their deltas forward and converting water into land. Scores of old lakes scattered over the county have thus been filled, and the upper surface is usually occupied by peat growth. The largest is Burton Moss, west of Burton-in-Kendal.

The Kent estuary has been silted up for miles from the head. There are three tracts of deposit. From Underbarrow southwards we have the oldest deposits, now largely covered with a rich peat which gives rise to a black fertile soil. South of this are the reclaimed marshes, which are still occupied by a growth of salt-marsh plants, and lastly we have the sand-banks at the mouth of the estuary, still covered by the tide at high water.

Looking over these flats from many points we get a

combination of three types of scenery, that of the mountain limestone hills, with the marshes below, and the slate hills beyond. Such views are specially fine. Prominent even among these are the prospects from Farleton Knott, Arnside Knott, and Whitbarrow, and from the fells around Helsington church.

We may now refer briefly to some of the minor features.

The influence of the lakes and tarns upon the scenery need merely be noticed in passing. We have said much of these sheets of water in earlier sections.

The accumulations of loose blocks detached by the weather from the cliffs above to settle in fan-like forms on the slopes beneath have a considerable effect upon the character of the hill-sides. These "screes," as they are termed, are more extensively developed in Cumberland, but they are quite abundant in Westmorland also. They are well seen on the lower slopes of the Langdale Pikes in the figure on p. 55.

Many gorges have been hollowed out by stream action since the occupation of the county by ice, especially along the principal joints and the belts of smashed rock among the hills of the volcanic group of rocks. The finest of these are Hell Gill and Crinkle Gill on the southern slopes of Crinkle Crags. They often contain waterfalls, as Dungeon Gill.

The waterfalls of Westmorland are remarkable for their beauty rather than for their size. Some of them occur in the gorges as just noted, others are rather cascades where the waters from the upland valleys pour down into the main stream; such are Sour Milk Gill from Easedale Tarn, Stock Gill Force at Ambleside, and Mill Beck from Stickle Tarn, Langdale. Others again, though but rarely, occur where hard rocks are in contact with others of less hardness along the courses of the streams occupying the larger valleys, as the small Skelwith Force in the Brathay and the more important Cauldron Snout on the Tees. The latter is due to the softer limestones having been worn away by the river down-stream, while the hard " whin sill " up-stream has resisted the wear.

The precipices of the district are small. The principal are found in the case of the lava and very hard ashes of the volcanic group, and in the mountain limestone fells. The hard well-jointed " whin sill " is responsible for the remarkable semi-circular cliff which forms the top of High Cup Gill.

In the limestone tracts some remarkable features are caused by the tendency of water to dissolve the lime. The formation of clints has already been noticed. Where streamlets run from overlying clayey ground on to the bare limestone, they sink down the fissures and gradually wear away a funnel-shaped hollow at the surface, known as a " swallow-hole." These swallow-holes may be seen in scores along the slopes of the Pennine chain. The water which enters them flows through the joints and bedding-places of the limestone, and often excavates caverns along the latter. Such caverns are abundant in the adjoining limestone district of Yorkshire, but are few and unimportant in Westmorland. The best known is

Pate Hole near Asby, a village about five miles south of Appleby.

The influence of vegetation on the scenery will be noticed in the chapter treating of Natural History, and as for the atmospheric effects, one need only remark that the variability of the climate, which is sometimes treated as a matter of regret, is responsible for scenic effects which are far more beautiful than would be the case were the climatic conditions of a more settled character.

9. Natural History.

Botany and zoology are the sciences which treat of the world's flora and fauna, but the study of the distribution of plants and animals—where they are found and why—forms part of the domain of geography, for from these we learn many facts concerning the past history of the land. Every one knows by sight a certain number of the plants and animals of his own county, and this knowledge will enable him to get some idea of the way in which their geographical distribution is effected.

Let us in the first place consider the plants of the county. Some of these are commonest in the south of England, others in the east, and others again in Scotland, while a very large number of the whole are spread over the entire island, and a few are very local, so far as our country is concerned.

These plants have not originated where they now grow. We have seen in the geological section that the district was once occupied largely by ice. At that time a few plants may have lived on the rocks of the higher fells, just as they do on the hills appearing above the ice of Greenland at the present day. But as the ice receded from the district tracts must have been left bare on which plants gradually sprang up, as their seeds were wafted by the wind, or brought by birds, or in some other manner, from other regions.

It is not to be supposed that the plants which were thus brought came from the regions above mentioned where they are now commonest, any more than that the Celtic speaking people of Britain who are now confined to the western high grounds entered from the west. There are many facts which shew that at no remote geological date, though before the beginning of historic times, England was joined to the continent along the tract now occupied by the Straits of Dover. This would form a ready route along which the plants which undoubtedly reached England from the continent could gradually migrate, just as, at a later period, successive immigrations of people came along that route, having only to cross the narrow straits. And as the more barbarous people were driven into the mountain fastnesses by their more highly civilised successors, so might the early plants be replaced by others which, under altered climatic conditions, were able to flourish.

But not all the human immigrants into Britain came by way of the narrow passage of the Straits of Dover. The sea-faring Danes and Norsemen, for instance, landed sometimes on the north-east and even on the west coast of England. Similarly some of our plants may have come in along some other route when England was united to the continent not merely along the tract now occupied by the straits, but by land masses which once existed over part of the site of the North Sea.

Certain plants now well established have been introduced by man. Most noticeable among these are such as grow in cornfields which have been accidentally brought into the country together with the corn. Some have been recently introduced and are not yet established, as the small toad-flax, others like the large blue speedwell (*Veronica Buxbaumii*), though of recent introduction, are now thoroughly established, and some, like the blue cornflower, have been so long inhabitants of the country that the period of their introduction is unknown.

It will be seen from these remarks that the question as to the mode in which our county became stocked with its plants is very complicated, and as it requires much knowledge of science to sift the evidence, this part of the study of distribution is only for those who are possessed of much botanical knowledge.

There are other facts connected with distribution however which can readily be tested. It will soon be found that the plants of Westmorland do not flourish equally in all parts of the county, for instance, those growing on the flats near the sea at Arnside are very different from those which live on the higher parts of Helvellyn. Very little observation will shew that there are two important causes of this difference among the plants of various tracts of the county, namely height above sea-level, and difference of soil. Let us first regard the difference of altitude. Many plants are confined to tracts less than 900 feet above sea-level, of which the gorse or whin is a good example. Above 1800 feet the bracken practically ceases. In the belt between 1800 and 2700 feet we find a remarkable assemblage of plants of an alpine character, such as the kidney-leaved sorrel (*Oxyria reniformis*), the rose-root (*Sedum rhodiola*), the purple-flowered saxifrage (*Saxifraga oppositifolia*), and the alpine rue (*Thalictrum alpinum*). Finally, above the last-mentioned height we get the most arctic of all the Westmorland plants—a little creeping willow (*Salix herbacea*) which, though occasionally found, it is true, at a somewhat lower level, as on Kidsty Pike near Haweswater, is on the whole characteristic of the hills above 2700 feet. It will be a useful exercise for the student to discover for himself what are the upper limits of the various plants with which he is acquainted, though he must be prepared to find an occasional straggler above the height to which the species as a whole ascends.

An easier study is that of the distribution of plants according to the soil, it being remembered that this soil in many cases varies in character according to the nature of the rock beneath, though some soils, like those formed of peat, are largely independent of the underlying rock.

Some plants are confined to the muddy silt of the salt-marshes by the sea-shore. A conspicuous example is the purple sea-aster or starwort with yellow eye (*Aster tripolium*), which grows on the salt-marshes north of Arnside.

The bog-plants live in bogs at various heights from

sea-level to the tops of some of the highest hills. Near sea-level we may yet meet with the Osmunda or royal fern, though it is less common than formerly owing to ruthless destruction by collectors. Even yet in the adjoining county of Cumberland it is in places sufficiently plentiful for use as litter for cattle. Other interesting bog-plants are the louseworts, the insect-eating plants known as butterwort and sundew, the beautiful little mealy primrose—the "bonny bird een" of Westmorland (*Primula farinosa*)—and handsomest perhaps of all, the grass of Parnassus (*Parnassia palustris*). In the pools among the bogs we find other plants, as the bladderworts and the pale blue-flowered water lobelia.

Some plants are confined to the rich soils along beck sides, like the globe-flower and the yellow balsam, the latter a truly local plant.

In the rough pastures we may observe many kinds of orchis, and along the hedgerows one often comes across two conspicuous plants of Scotch type, namely the melancholy thistle (*Carduus heterophyllus*), and the great bell-flower (*Campanula latifolia*).

In the limestone district is a group of plants which flourish notwithstanding the general dryness of this tract. Among such are the centaury, the rock-rose, and the lady's finger. But the most noticeable plants of the limestone tracts are those which live in the fissures of the cliffs and clints, such as the harts-tongue and green spleenwort ferns, and the yew.

The determination of plants growing on different kinds of soil will be an interesting study. It will sometimes be found that the same plant has a different growth under different conditions, thus the golden-rod of the lowlands is not quite like that found in the fells, and the vernal whitlow-grass (*Draba verna*) has a different growth on limestone cliffs to that which it exhibits on the slate rocks.

Before leaving the consideration of the distribution of plants there is one matter concerning which a few words must be said.

We saw that at over 1800 feet a number of plants were found which, with us, do not occur below that height. These however are widely scattered in European mountain regions, many being found on the Scotch hills, the Alps of Switzerland, the mountains of Norway, and on lower ground within the arctic circle. They are at the present day mainly characteristic of alpine and arctic regions, and it is believed that they became established in our country during the glacial period, occupying then the British lowlands, just as they now live on the lowlands within the arctic circle. As the climate grew warmer they were displaced by other plants which were able to flourish to so great an extent as to exclude these "alpines," which accordingly were driven higher and higher, and are now found obtaining here and there a precarious footing upon our higher fells, from which perhaps they are doomed to disappear at no distant date. Let us hope that the disappearance, if it comes, will be natural, and not quickened by the wanton removal of the roots of the plants by the too eager collector.

In a county of which the scenery has within recent times had a marked effect upon the dwellers therein, a few words may be added as to the effect of the plants upon that scenery.

Many plants grow in sufficient number to produce a striking influence upon the view. The flowers of the rag-wort in the rough pastures, the curious growth of the cotton-grass when in seed, and the leaves of the alpine lady's-mantle, which, growing in masses on the rock ledges of the fell-sides, sometimes gleams with a green of almost metallic sheen, may be cited as examples. There are, however, two plants whose influence is particularly pronounced, namely the heather and bracken. The effects of heather are most striking on the moors on either side of the Eden valley. The amount of heather in the Lakeland portion of Westmorland is comparatively small, but it is in this tract that the effects of the bracken are so fine. Of it Wordsworth speaks thus :—"About the first week in October, the rich green, which prevailed through the whole summer, is usually passed away. The brilliant and various colours of the fern are then in harmony with the autumnal woods : bright yellow or lemon colour, at the base of the mountains, melting gradually, through orange, to a dark russet brown towards the summits, where the plant, being more exposed to the weather, is in a more advanced state of decay."

About the animals of the district we need say little. Gifted with power of locomotion, their distribution is as a whole wider than that of the plants.

The mammals have suffered much from the hands of man, and especially is this the case with the larger beasts. The wild cat, the wild boar, and the badger are now extinct, though their previous occurrence is indicated by some of the place-names. There are several Grisedales in the Lake district, one, near Ullswater, being in our county. These dales are so named from "gris," an old name for the wild boar. Again, the word Brockstone means "the stone of the brock" or badger. The fox is yet with us, and a fell fox-hunt conducted on foot is still an exhilarating pursuit.

Birds are abundant enough in the lowlands, though on the fells there is a singular lack of bird-notes. The call of the wheatear is an exception, and we may often hear the cry of one or other of the species of hawk, and more rarely the hoarse croak of the raven. The grouse-moors of the county lie chiefly in the eastern part, on Shap Fells, and on the east side of the Pennines north-west of Stainmore. The red grouse is confined to Britain.

There is little of general interest with regard to the distribution of the reptiles and amphibians, but the fishes of the county present some noteworthy features. Char are found in Windermere and in smaller quantity in Haweswater and Ullswater, and in Ullswater is the fish known as the schelly. In Britain both char and schelly are characteristic of the lakes of the hill-regions.

So far back as the beginning of the eighteenth century we learn that char were "baked in pots, well seasoned with spices, and sent up to London as a great rarity," and a few still find their way thither in the same manner.

Of the invertebrate creatures much could be written in detail, but it would require considerable knowledge of zoology. Leaving the mass of these animals unnoticed we may refer to one, the mountain-ringlet butterfly (*Erebia cassiope*), which is abundant at a height of over two thousand feet on the fells, and though found in Scotland cannot be met with between here and Switzerland. Like the alpine plants it is probably a survival from the organisms which spread over our country during the Glacial Period.

10. Climate.

The climate of a country is the result of the combined effect of the different variations of what is commonly termed the weather. The most important factors in determining the climate are temperature and rainfall.

The great variations in the climate of the world depend mainly upon differences of latitude; thus we speak of tropical, temperate, and arctic climates; that of our country being temperate. Another important factor in controlling climate over wide tracts of country is nearness to the sea, so that along any great climatic belt we have variations according to the geographical conditions, the extremes being "continental climates" in the centres of continents far from the oceans, and "insular climates" in tracts surrounded by ocean. The continental climates are marked by great variations in the seasonal temperatures, the winters tending to be exceptionally cold and the summers exceptionally warm, whereas the climate of many insular tracts, including Britain, is characterised by equableness and by mild winters and fairly cool summers.

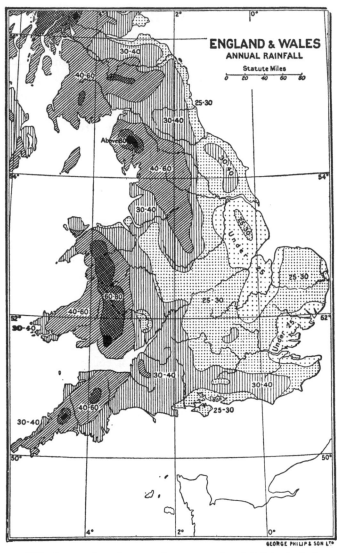

(*The figures give the approximate annual rainfall in inches.*)

Again, an insular climate tends to be more humid than a continental climate. Great Britain, then, possesses a temperate insular climate.

Different parts of England possess different climates, and we must now consider wherein and why the climate of Westmorland varies from that of other parts of the country.

Two especially important points must be regarded in contrasting the climatic conditions of Westmorland with those of other parts of England. Firstly, Westmorland is further from the European continent and nearer to the Atlantic Ocean than is the eastern portion of England, and its climate therefore departs more widely from the continental type than does that of eastern England. The nearness of the ocean in addition to moderating the temperature, as stated above, greatly affects the rainfall, as will be seen by studying the rainfall-map, where it will be noticed that the western side of our island on which Westmorland lies is much wetter than the east side, for the vapour-laden winds from the Atlantic deposit much of their moisture on the west, and blow eastward as drier winds. In the second place the Westmorland climate is largely influenced by the great amount of elevated land within the county boundaries.

As the evaporation of water and its subsequent pre-cipitation as rain is dependent upon changes of tempera-ture, we may consider first the temperature conditions.

England and Wales are situated in a belt having a mean annual temperature of about 50° Fahr., the mean temperature for January being about 40° Fahr. and that for July 60° Fahr. and these figures hold good for West-morland, whereas in East Anglia the January and July temperatures are about 38° and 62° and in parts of western Ireland about 42° and 58° respectively. It will be seen then that, comparing summer and winter temperatures, East Anglia has a less equable, and western Ireland a more equable climate than Westmorland.

This distribution of temperature shews that latitude alone does not produce the variations, otherwise it should be colder as one passes northward. It has long been known that temperature variations in our island are greatly affected by the prevalent south-westerly winds bringing heat from the waters of the Atlantic. These waters off our coasts are exceptionally warm for their latitude, owing to their movement from the warmer south-westerly seas towards our shores on the north-east. This movement is that of the Gulf Stream, a drift of the surface-waters of the Atlantic in a north-easterly direction caused by the prevalent winds.

It is impossible here to discuss the principles which control weather changes. It must suffice to say that our weather is largely influenced by the prevalence of *cyclones* from the Atlantic. The air movements are *cyclonic* or *anticyclonic*. In a flowing stream you may often observe a chain of eddies bounded on either side by more gently moving water. Regarding the general north-easterly moving air from the Atlantic as such a stream, a chain of eddies may be developed in a belt parallel with its general line of movement. This belt of eddies or cyclones as they are termed tends to shift its position, sometimes passing over our islands, at others to the northwards and at others again to the southwards. To the shifting of this belt most of our weather changes are due.

When the country is influenced by a cyclone it is often windy, while when under the influence of an anticyclone it will more probably be still and dry. Cyclones, then, are apt to be accompanied by wind and rain, anticyclones by calm, during which there may be bright sunshine with warmth in summer, clear cold weather in winter, and fog in autumn.

There is one period of the year when the distribution of the winds in our country is affected in a different way by the temperature of the great continental mass to the east. The conditions are then such that the belt of cyclones is, as it were, pushed back over the ocean, and we experience the east winds which are often prevalent during the month of March.

Let us now further consider the rainfall. Cold air can hold less water vapour than hot air, and accordingly when the air rises and becomes chilled in the higher parts of the atmosphere it tends to part with its moisture as rain. This air may rise by expansion, which makes it lighter, or by blowing up a rising land-surface. The importance of the latter cause is great, as may be seen by studying the map of rain-distribution in our island, when it will be noticed that the areas of high rainfall coincide with the elevated tracts. The large amount of rain which falls in Westmorland is mainly due to the vapour-laden winds from the Atlantic being forced up the hills and precipitating their moisture, and accordingly the greatest amount of rainfall occurs practically on the tops of the ridges which face the ocean. The greatest rainfall in the county is found in the westerly tract of high ground, where it is over 80 inches per annum. It decreases eastward, and sinks to under 40 inches in the Eden valley. The rainfall of the great Pennine ridge east of that valley has not been studied in detail. It may be remarked by contrast that the driest parts of England have less than 20 inches per annum.

The figures given as to the annual rainfall in the Lake District in some books are very misleading, and Seathwaite in Borrowdale (Cumberland) has an unenviable reputation for an average rainfall of 154 inches per annum measured in six years. But this is exceptional. It must be remembered also that the amount of rainfall does not give a measure of the length of the period during which it falls. The very heavy falls in parts of Westmorland cause a much greater amount to be precipitated in a given time than in some other tracts where the fall is gentler.

The amount of sunshine recorded differs for different parts of England, the greatest amount being in the south, and the least in the southern part of the Pennines. Along the greater part of the south coast more than 1700 hours per annum are recorded. The smallest amount for England is under 1200 hours. The greater part of Westmorland lies in a belt receiving more than 1200 and less than 1400 hours.

From the foregoing remarks it will be gathered that the bad reputation of the district is not altogether deserved, and in the summer months, during which visitors arrive in greatest number, the number of hours of rainfall is not excessive.

One or two matters of detail in conexion with the Westmorland climate should be noticed.

As regards wind the valleys of the upland tracts are comparatively sheltered and violent gales are not frequent, though the wind is often very strong on the ridges. One local phenomenon deserves mention, namely the " Helm wind,"—so called because it is accompanied by a helmet-shaped cap of cloud on the Pennine ridge. It is an east wind, and is most pronounced during the prevalence of the early spring east winds. Leaving subsidiary features out of account, the wind does not differ from winds in other parts of the world which have a similar configuration. The Pennine ridge, as we have seen, has a gentle slope eastward and a steep scarp facing west. The air rises slowly up the gentle eastern slope and rushes violently down the western scarp, and its moisture becomes condensed on the summit ridge.

Severe frost is not so frequent in Westmorland as it is in parts of south-eastern England, where the average winter temperature is lower. Snow falls in the winter season on the higher fells and often lies long there, but there is no very great amount of snow in the lower regions. Some of the most severe snow-falls occur during the prevalence of east winds, and therefore tend to be heavier on the Pennine ridge than upon the hills of the Lake District.

11. People—Race. Language. Settlements. Population.

When the Romans invaded our island, it was occupied by people whom we are accustomed to speak of popularly as the early Britons. These people, however, were not all of one race, and we may briefly consider who they were. Of the earliest inhabitants of our land known as the " Palaeolithic " men we have no trace save the implements which they left behind, and of these none have been found in Westmorland. Long after the disappearance of these people, a short swarthy race arrived from the continent and spread widely over Britain, certainly penetrating into Westmorland, as indicated by their relics. Whence they came and who they were we know not for certain ; all we can say is that they were an earlier set of immigrants than the Celts who succeeded them.

These early men were displaced by a taller and more powerful people armed with better weapons, who, however, probably did not completely destroy their conquered enemies, but held the survivors in bondage as slaves. The more powerful race, the Celts, are supposed to have come into Britain at two distinct times. The earlier immigration was of a Celtic race who spoke a language like the modern Gaelic : these people are known as the Goidels or Gaels. Subsequently another Celtic race—the Brythons— speaking a language like Welsh arrived.

There were probably, then, in our county, even if palaeolithic man never arrived there, three races before the Roman invasion, one pre-Celtic and two Celtic, though some believe that the inhabitants of what is now Westmorland were essentially Goidels.   Be this as it may, at the time of the arrival of the Romans, the north of England, including Westmorland, was occupied by a powerful Celtic tribe, that of the Brigantes.   This tribe was divided into sub-tribes, but their distribution is uncertain, and it will be sufficient for our purpose to know that the people of the Brigantes in those days inhabited what is now Westmorland.   It is doubtful whether any important town of the Brigantes lay in Westmorland territory.   In all probability the inhabitants were scattered over the area, living in small hamlets.   It is also doubtful whether any traces of the Brigantes can be found among the characteristics of the existing people of Westmorland.   But many of their place-names, as subsequently remarked, still survive.

In the first century A.D. the influence of the Romans began to be felt, and was exerted in Westmorland for nearly four hundred years.   Important as was the civilising influence of the Romans upon the inhabitants, as we shall see in a later chapter when we come to speak of the Roman roads, the Roman occupation produced little permanent effect upon the physical characters and the language of the inhabitants.   The occupation was essentially military, and the Roman legions were composed of a soldiery of mixed race gathered by the Romans from various quarters of Europe.   During the temporary wane of Roman influence in the fourth century and after the final independence of Britain from the Roman yoke in the following century there were invasions of the district from the north by the Picts and Scots, but these appear to have been of the nature of raids, and to have had little effect upon the character of the modern inhabitants.

At the beginning of the seventh century the Anglo-Saxons began to occupy the district, entering from the south and also over the Pennine passes. These immigrants were mainly Anglians.

In the ninth century the Danes came into the district from the east, over the Pennine passes, and in the tenth century the Norsemen, who had previously descended upon the Isle of Man, came over the sea to the west coast and effected a permanent settlement in the district.

In 1092, William Rufus brought an army to the north and the Norman settlement began. This was the last important immigration of the various races into what is now Westmorland. Of these invasions that of the Romans had a striking effect upon the civilisation of the people ; the Anglo-Saxon invasions of England gave us our language (afterwards modified by Norman influence); while the present physical characters of the inhabitants of Westmorland are with good reason considered to be chiefly due to the incoming of the Norsemen from the Isle of Man.

It has been well said that the history of a country is written on the face of the country itself—in the names of its towns and villages, its rivers, mountains, and lakes. And so we shall find that the Westmorland place-names give much evidence of the character of the different invasions. As the later invaders naturally occupied the fertile lowlands rather than the barren fastnesses of the hill tracts, we find a number of place-names of British origin in the upland regions, and accordingly they are more frequent in the elevated tracts of West Cumberland than in Westmorland.   The word "combe" (Welsh *cwm*), for example, used to indicate a half-bowl shaped hollow in the hills, is still in general use.

Pavey Ark: a combe on Langdale Pikes

The occupation of the southern lowlands by the Saxons is shewn by the number of *hams* (*ham* = a home) in that tract, such as Heversham and Beetham, while the Danish word *by* (a village) is found abundantly in the Eden valley, where settlements were made by the Danes who came over the Pennines.   Such are Melmerby, Gamlesby, Ouseby, Appleby, and a large number of others, though as the word is also Norse it may well be that many such names were given by the Norsemen.

M. W.   6

A large number of these Norse and Danish words occur in place-names like Ambleside, Milnthorpe, Murthwaite (*side* = settlement ; *thorpe* = hamlet ; *thwaite* = clearing). Several of them are still in use in the local dialect for physical features, as *foss* or *force* for a waterfall, *gill* for a stream, and *fell* for a hill, while yet others are in use for other things, thus a *gimmer-lamb* is a female lamb (Danish *Gimmerlam*), and *smit* is used for the smear of colour with which sheep are marked.

Traces of settlements of the pre-Roman inhabitants have been found in many places, often on high ground. The Anglian, Saxon, Danish, and Norse immigrants probably occupied some of the settlements which had been founded by their predecessors, but there is no doubt that they founded many new hamlets and villages. It has been seen that they occupied at first the richer lowlands, and in course of time no doubt took into cultivation the fertile floors of the upland valleys, gradually extending up these valleys into the inner recesses of the fell country.

On the arrival of the Normans the county was parcelled out into large areas divided among the Norman barons, under whom the general mass of the inhabitants lived as bondsmen ; but the Norman influence was mainly restricted to the lowlands of the south and north-east of what is now Westmorland, and those who had penetrated into the heart of the dales remained undisturbed in the possession of their small freehold estates. Thus arose an independent set of small farmers known as estatemen, or "statesmen," having properties of from thirty to three hundred acres, which descended to the eldest child. The farm buildings were usually situated in the valleys, surrounded by fields in which oats and other crops were cultivated, and in which cattle grazed. These fields extended up the hill sides, and far above was the open fell on which the sheep grazed, and from the peat-mosses of which peat was dug for fuel. It is only within the last half-century that these statesmen have practically disap-

A Westmorland Farm

peared, and the influence of their independence as keeping up and intensifying the sturdiness due largely to the Norse blood is undeniable. The stone walls which are so marked a feature in a Westmorland dale prospect are chiefly due to the separation of the various farm-lands and the subdivision of each farm for the different uses to which its portions were put. The picture of Langdale (p. 55)

6—2

gives an idea of these Westmorland farms with their walls running up the fell sides.

Owing to the physical character of the country and the absence of any great industrial centres the number of people living in the county is small. According to the census of 1901 the population was 64,409, being the smallest population of any English county except Huntingdonshire and Rutland. But even the two counties mentioned have a denser population than has Westmorland, which is the only one in England with a population of less than 100 per square mile. It is interesting to notice that the neighbouring county of Lancashire and the county of Middlesex are the most densely populated, having more than 2000 people to the square mile.

In every county in England, with the exception of four, we find that the population is now increasing. In Westmorland it is diminishing. The increase elsewhere is owing to the growth of many large provincial towns consequent upon the growth of their industries. But in Westmorland the towns are small and have no great or special industries, and hence they have not been materially affected by the modern industrial revolution.

12. Agriculture.

As the inhabitants of Westmorland have from time immemorial essentially subsisted by agriculture, we must devote some attention to its consideration.

In doing so, it is well to note first the principal factors which control the agricultural operations of any given county or district. These are latitude, altitude, climate, soil, character of the people, and outlet for surplus produce.

Latitude is of special importance in affecting the nature of the corn crops, for the county is too far north to allow of successful cultivation of wheat on a large scale, while on the contrary it suits the growth of oats, and accordingly we find that this is the chief kind of corn grown.

Altitude is responsible for the dividing line between the areas devoted to the cultivation of crops and the permanent pasture. The area of the former is chiefly in the lowlands and valley bottoms, whereas much of the latter is on the upper portions of the fells. The line separating the tracts where arable cultivation is profitable from those of permanent pasture may in the Lake District be roughly placed at a height of 900 feet above sea-level.

It will have been gathered from the remarks made in a previous chapter that the county enjoys an equable climate with fairly warm winters and cool summers, and that it is essentially humid. Such a climate is very favourable to the growth of many crops, and especially of root crops.

We have considered the soil in the geological chapter. It was there seen that there were four important varieties, namely the clayey soil of the slate-rocks, the thin light soil of the Mountain Limestone tracts, the sandy soil of the New Red Sandstone region, and the mixture of peat and silt in the estuaries and infilled lakes of the northern part of the county. The agriculture of each of these soils is marked by special features.

The character of the inhabitants is also of great

importance. The physical strength, capacity for work, and tenacity of purpose inherited from their Scandinavian ancestors and fostered by the struggle against physical conditions which cause the produce to be wrung from the earth by hard labour, have enabled the natives to overcome the difficulties with which they were confronted, when a weaker and more indolent race would have been worsted in the struggle.

Lastly there is the outside market for the produce. In the early days corn and roots were grown in sufficient quantity for home consumption only, and the same may be said of the supply of meat. The wool of the sheep alone was carried to the markets to exchange for such necessaries of life as could not be obtained on the spot. Of recent years, owing to the expansion of large towns outside the district and the facilities for transport caused by the development of the railways, the agricultural produce of the county is taken further afield, and accordingly many cattle are reared, especially dairy shorthorns. As the result of the rearing of this stock much milk and butter are sent to various large towns in southern Scotland and the north of England, and dairy stores are sent to places in all parts of the kingdom.

The following figures from the journal of the Royal Agricultural Society give the acreage devoted to the different corn crops and to live stock in 1907.

Corn Crops				Acreage
Wheat	...	...	...	127
Barley	...	...	...	442
Oats	...	...	...	13,791

Live Stock				*Numbers*
Cattle	...	...	...	70,647
Sheep	...	...	...	415,360
Pigs	...	...	...	4,729

From this it will be seen that rather more than four-fifths of the entire area of the county are devoted to sheep pasture and only $\frac{1}{35}$th to the cultivation of corn, which is almost exclusively oats.

Herdwick Sheep

Much of the sheep pasture, as already stated, is on the higher hills, which are chiefly occupied by the old slate rocks, though parts of the mountain limestone tract with its sweet herbage also furnish excellent pasture.

The sheep of the county were at one time largely of

the Herdwick breed, of which tradition says that they
escaped from a vessel of the Spanish Armada wrecked on
the west coast of Cumberland and thence spread through
the district. These Herdwick sheep are now fast dis-
appearing, owing in great part to the destructive " fluke "
worm, and are being replaced by Cheviot and Scotch
crosses and by a breed of long-woolled sheep of Lincoln
origin, now known as Westmorland Long-wools.

Damson Blossom, Winster Valley

The cattle are mainly reared in the lowlands, which,
as we have seen, occur along the Eden and its tributaries
to the north, and in the neighbourhood of the Kent
estuary in the south.

Some fruit is grown in the lowlands. The damson-
trees of the Winster valley are shewn in the picture.

In the lowlands also a certain amount of potatoes are cultivated, and, thanks to the humid climate, other root crops are extensively grown. The swedes of Westmorland are the finest in England, and the growth of mangolds is coming in.

But, as will be gathered from the table of acreage above given, by far the most important crop of the county is oats, which are grown not only in the lowlands but in the lower parts of the valleys of the higher fell tracts.

13. Industries.

The inhabitants of the county having been from time immemorial a pastoral and agricultural people, there is little to say about their other industries, save mining and quarrying, which will be considered in a separate section.

At the present time there is no centre of industry except Kendal, and here the one great industry of former times has given way to several minor ones. The manufacture of woollen cloth was introduced into this town by immigrants from the Low Countries in the time of Edward III and flourished for several centuries. The well-known "Kendal green" cloth was largely used for making clothes. The green colour was derived from the Dyer's broom (*Genista tinctoria*) which grows in the neighbourhood. This industry has disappeared. Now such varied articles as blankets, carpets, boots, combs, and (in the neighbouring village of Burnside) paper are manufactured.

But though there is now no great industry in the county, there are several old manufactures of considerable interest. An attempt was made by Ruskin to revive the old linen industry, and for a few years hand-woven linen was made in Langdale, but the revival was not successful and the manufacture has now been removed from Langdale to Coniston in the neighbouring county.

Another industry was the making of charcoal. The copses of the lowlands furnished excellent wood for it, and charcoal-burners were found scattered about the county. From this another industry sprang up, which still survives, namely the manufacture of gunpowder, which is carried on at Elterwater and near Holme.

The wood of the copses was also used for making cotton-reels or bobbins, and bobbin-mills existed at Staveley and at Troutbeck Bridge near Windermere. The copse-woods also furnished the material for the manufacture of baskets.

In addition to agriculture, mining, and quarrying, a most important industry in the county is provision for the tourist and visitor. Large hotels have sprung up on the shores of Windermere and Grasmere, and many smaller ones are found in the dales of that part of Westmorland which forms a portion of the Lake District. Lodging-houses also cluster thickly in many places, as Windermere village, Bowness, Ambleside, and Grasmere. Many people are, moreover, employed in shopkeeping mainly for the visitors, and in their conveyance from place to place.

The conversion of the Lake District into a tourist centre has indeed produced a most profound effect upon the inhabitants of the county. To realise this let the visitor take his stand at the end of the pier at Waterhead, Windermere, when a steamer arrives, and see the crowd of vehicles which soon depart laden with visitors to different parts of the district.

14. Mining and Quarrying.

The principal metalliferous ore which has been worked in past times in Westmorland, and is still being worked, is lead-ore. The chief mines of this ore occur in the slate series, in the Borrowdale volcanic rocks of Glenridding, near the head of Ullswater. In addition to the lead, much silver has been extracted from this ore. Veins of lead-ore have also been worked in the past on the hills between Kentmere and Long Sleddale, in the Upper Slate rocks.

The Carboniferous rocks of the Pennine chain have long been famed for the rich veins of lead-ore which they contain. The principal mines along this chain in Westmorland were situated on the hills behind the hamlets of Dufton, Murton, and Hilton.

Of non-metallic minerals, a little barytes is mined in the Carboniferous rocks of Hilton Beck, and some gypsum in the New Red Sandstone of the Eden valley.

It has already been noted that the productive coal-measures are practically absent from Westmorland, but thin seams of coal occur in lower members of the Carboniferous strata, and have in former times been worked for

A Slate Quarry, Langdale

local use about Reagill near Shap, near Stainmore, and in Mallerstang.

Abundant material for building-stone is found in the county. The Volcanic series furnishes compact green-stones which are largely used in that part of the county which is occupied by these rocks. The gritty beds and some of the coarser slates of the Upper Slate rocks are likewise largely used for houses. The mountain limestone furnishes building-stones also, and much of the town of Kendal is built from this rock, obtained from the neigh-bouring quarries. Some of the Carboniferous sandstones are also used in places. Two important red sandstones are found in the New Red Sandstone rocks of the Eden valley, and the towns of Appleby and Kirkby Stephen are mainly built of these materials, and, in the case of the latter town, of another stone of the New Red Sandstone called brockram. Slate is obtained from two bands in the slate rocks. One of these is found in the Volcanic series and crosses part of the county from Mosedale, a few miles south-east of Shap, to Little Langdale, where it enters Lancashire. This slate is of a green colour and is very good. It is largely exported in addition to its being used in the district. The largest quarries in the county are in Langdale, Troutbeck, and Kentmere. The second band, of a softer nature and of a grey-blue colour, occurs in the Upper Slates and may be traced from Long Sleddale to near the head of Windermere, being parallel to the former band. The largest quarries in the county for this slate are on Applethwaite Common east of Troutbeck. Fissile sandstones are found in the Carboniferous rocks near Shap and Kirkby Lonsdale, and are locally used for roofing and flagstones.

A former industry now extinct was the manufacture of slate-pencils from the soft slates of the Skiddaw slate group found in the streams west of Shap.

Of ornamental stones, the best known is the granite of

Peat-digging, Lyth

Shap Fells, a grey or pink rock with large rectangular crystals of pink felspar. It has been largely worked at different times and when polished is used for pillars and tombstones. It has been extensively used for decorative purposes in many parts of England, and even exported to the continent. A limestone veined with red has been worked for "marble" in the neighbourhood of Kendal, and other false marbles have been extracted from the Carboniferous limestone around Kirkby Lonsdale and elsewhere.

Road-metal is furnished by many of the harder rocks, and the materials of the immediate neighbourhood are generally used, but the "whin sill" rock has often been conveyed to some distance for this purpose.

Lime is burnt in many places where the Carboniferous limestone is developed, and a thin seam of limestone at the base of the Upper Slate series has also been used for this purpose.   Bricks and coarse pottery have been made from the boulder-clay near Clifton in the Eden valley.

Peat for fuel is dug from the turbaries or peat-bogs on the higher hills in many places, and also in the flat mosses which are found in the low ground of the Kent estuary and in the neighbourhood.

15. History of Westmorland.

The Roman occupation of Britain began with the visits of Julius Caesar in B.C. 55 and 54, but the Romans first arrived in what is now Westmorland territory under Agricola in A.D. 79.   With that event the history of Westmorland begins, for earlier events occurred in pre-historic times.   For nearly four centuries they dominated the district, with occasional raids by the Picts and Scots, and an insurrection of the Brigantes which was soon quelled.

The Romans first arrived from the south along the Lancashire plain, and subsequently from the east over

Stainmore. They opened up the country by constructing an elaborate series of roads which will be described in a subsequent chapter, and protected them by military camps at points of strategic importance, and especially under Agricola the Brigantes were introduced to the civilisation and luxury of the Roman conquerors.

In the fourth century the decline of the Roman Empire had begun, and at the beginning of the fifth century the Emperor Honorius gave the Britons their independence.

Shortly after the departure of the Romans, who were gradually leaving the country between 410 and 430, the English invasion began by the arrival of Teutonic peoples from the country near the Elbe. There were three Teutonic tribes dwelling near the Elbe, namely the Saxons around the mouth of that river, the Angles further north, and still further north the Jutes. All these people have been termed Anglo-Saxon or English, and as they conquered portions of our county it in turn became English. By the end of the sixth century they had annexed the lowlands of England, the Jutes occupying Kent, the Saxons a tract north and west of Kent, and the Angles the centre, east, and north-east of England.

All this time the Westmorland area remained part of an independent British kingdom extending from the Dee to the Clyde, and bounded on the east by the Pennine hills. This kingdom was called Strathclyde and was divided into a series of smaller States, one of which—Cumbria—included what is now Cumberland, Westmorland, and North Lancashire. We know little of what

went on in this region between the departure of the
Romans and the arrival of the English, Danes, and Norsemen, but there were in all probability dissensions among
the different States.

After the battle of Chester in 607, when Ethelfrith
king of Northumbria defeated the British, the portion of
Westmorland south of Shap Fells was occupied by the
Angles and became English while the northern portion
remained as a part of the British kingdom of Cumbria,
which was only attached to the English kingdom. The
Angles from Northumbria had in the meantime come
over the Pennines into northern Westmorland, and a
Saxon invasion occurred in the south.

Again, the Danes came over the Pennines into the
northern part of the tract in the ninth century and the
Norsemen arrived from the west in the following century.

The details of these various invasions of the northern
part of the county are veiled in obscurity. Mr Burton
writes: "Of these territories it can only be said, that at
this period, and for long afterwards, they formed the
theatre of miscellaneous confused conflicts, in which the
Saxons, the Scots, and the Norsemen in turn partake.
Over and over again we hear that the district is swept by
the Saxon king's armies, but it did not become a part of
England until after the Norman conquest."

In 945 Cumbria was conquered by the Scots and
became part of the kingdom of Scotland, while what is
now southern Westmorland still belonged to England.

In 1092, William Rufus came to the north, took
possession of the northern tract of Westmorland and

Cumberland, and established the present boundary between England and Scotland, and what is now Westmorland for the first time became wholly English.

Obscure as are all the events which happened between the departure of the Romans and the arrival of the Normans their influence upon the inhabitants of the area was most marked. The people of the Brigantes had disappeared and the inhabitants had acquired those Scandinavian

Kendal Castle

characters which they still possess, while the English language had become established.

After the conversion of the northern tract into English soil, the area which now comprises Westmorland was in two baronies, that of Kendal in the south and Appleby in the north.

It was not until the time of Henry I, as stated in the

first chapter, that the northern tract was separated from the land of Carlisle and joined to the southern tract to form the county of Westmorland.

After the death of Henry, Westmorland (or at any rate the northern part of it), together with Cumberland, was given up to the King of Scotland, but in the third year of the reign of Henry II, 1157, it was annexed to the English Crown and once again, and finally, became a part of England.

The subsequent history of the county is simple. In 1173 the King of Scotland destroyed the town of Appleby, and the Scotch invaders again burnt the town in 1388. When civil war broke out in the seventeenth century, Anne, Countess of Pembroke, garrisoned Appleby Castle for the King, but it surrendered to the Parliamentary army in 1648.

The last battle in Westmorland, and indeed on English soil, took place at Clifton, three miles south of Penrith, in 1745. The troops of the Pretender, Prince Charles, flying north after their defeat at Derby, came through the Shap Pass, were attacked by the Duke of Cumberland, and fled to Penrith.

Let us turn from warfare to more peaceful pursuits.

The lack of important events throughout the Middle Ages shews that the inhabitants of the county were left free to follow their own pursuits. During those ages the towns grew, the cultivation of the lowlands gradually crept upwards towards the valley-heads, and the people lived a simple life as farmers with no disturbances from the outside.

A most important circumstance which influenced the welfare of Westmorland was the development of the taste for the admiration of mountain-scenery among the English-speaking people, for not always has this taste been with us; indeed until comparatively recent times the contemplation of mountains filled the minds of most people with horror. The literature devoted to the admiration of the Lake District dates from the middle of the eighteenth century, and culminates in the writings of the group of poets of whom Wordsworth was the chief. Attracted by this literature, and all difficulties of access overcome by the introduction of railways, travellers turned their attention to the district in ever-increasing numbers.

16. Antiquities—Prehistoric, Roman.

Our knowledge of the history of the inhabitants of Westmorland is chiefly derived from a study of written records, but it is not entirely dependent on it, for the relics of the early inhabitants give us further information even about the periods of which records were made in writing.

The tract now forming the county was, however, inhabited in times earlier than those concerning which we have the first written records, and our knowledge of the state of the inhabitants of these early days is derived entirely from an examination of the relics left by them, either in the form of structures such as grave-mounds and stone-circles, or of various weapons and other articles which have resisted decay, and been preserved to the present day.

The periods previous to the first written records are usually spoken of as "Prehistoric," and we will now consider the nature of the remains which have come down to us from these times.

Before the use of metal for forming tools and weapons was discovered, these were chiefly of stone (and in some cases of bone), and we are enabled therefore to divide prehistoric times into the Stone Age and the Prehistoric

A Stone Implement from near Kendal
(*In Kendal Museum*)

Metal Age. We will begin with the earlier of these ages—that of stone.

Antiquaries have found that there are two very different classes of stone implements marking two quite distinct ages of civilisation, of which the later was far more advanced than the earlier. We may speak therefore of the Palaeolithic or Old Stone Age, and the Neolithic or New Stone Age. In the older age the implements were formed by chipping the stone into shape, and the art of grinding and polishing them was unknown. Instruments of this type are found in the river-gravels and caverns of the more southern parts of our island, but are unknown in the northerly tracts including Westmorland. We need not, therefore, dwell further upon the remains of the older Stone Age.

Several instruments of the later or neolithic age have been found in the county. Among them are many stones which have been chipped and ground into the form of a broad chisel with a sharp cutting edge at the broader end. These are often polished and were probably used as hatchets. A number of dumb-bell shaped stones have been found in the neighbourhood of the larger lakes which may have been used as net-sinkers by fishermen. Other types also occur. The implements are mostly made from some of the hard rocks of the district. It is not to be supposed that all the stone implements which have been discovered in the county belong to the Stone Age, for stone was used long after the introduction of metal, and has indeed been in use quite recently for some purposes, as for instance the "strike-a-light" used for igniting tinder.

Of structures left by these men of the Stone Age there are few in the county which were of certainty made by them. The chief are the "barrows" or burial-mounds. Of these barrows there are two types, known as long barrows and round barrows. The long barrows belong to the early neolithic period and the round barrows to a later period. One near Crosby Garrett has been examined, and the remains of several cremated bodies found in it, with bones of various animals, chiefly domesticated.

From study of the relics of the later Stone Age found in other places we know that the English dwellers of this age were hunters and fishermen, possessing domestic animals, and having some knowledge of agriculture. They were also acquainted with the art of making rude pottery.

The introduction of metal into the county was gradual, and long after objects formed of metal were introduced stone no doubt continued to be the principal material from which implements were fashioned. The first metal to be

A Bronze Implement found in Westmorland

used for the purpose was not iron but the alloy of copper and tin which we call bronze, for the art of smelting iron was much more difficult than that of making bronze. Accordingly the earlier prehistoric age of metal was a Bronze Age. At first the Bronze Age men imitated the stone implements which were in use, and a large number of the bronze instruments are more or less modified forms of the hatchet-shaped stone instrument, but the introduction of metal allowed of the formation of far greater variety of forms than could be fashioned in stone, and in our county, besides the hatchet-shaped forms, many bronze dagger-like weapons and swords have been discovered. Metal also lent itself to the making of ornaments, accordingly bronze bracelets and other ornaments have been found in the county.

The structures which have been referred to the men of

Stone Circle, Moor Divock

Westmorland living in the Bronze Age are very abundant. Numerous round barrows of this age are scattered over the county ; they were used as burial-mounds. Many of them are largely composed of stones, forming cairns, though these are commonly covered with soil overgrown with vegetation.

The county is peculiarly rich in so-called " Druidical

Circles," which, however, have nothing to do with the
Druids. They cluster thickly on Moor Divock near the
foot of Ullswater, and in the vicinity of Shap and Crosby
Ravensworth. They are usually formed of large stones
set upright in the form of a circle, though sometimes we
meet with double circles one within the other.

The remains of settlements of prehistoric tribes have

Stone Circle, Oddendale, near Shap

also been found in many places, though naturally little
more than the foundations. They occur around Shap,
Crosby Ravensworth, Windermere, and Kirkby Lonsdale,
and appear to have been groups of huts, often protected
by an enclosing wall. Various camps and earthworks
have been referred to these times.

There is evidence that iron had been introduced into

Britain before the arrival of the Romans in our island, but as no relics discovered in Westmorland can be definitely assigned to this period we may dismiss the subject with this brief notice, and proceed to a consideration of the relics of the Roman occupation of our land. In so doing we pass definitely from prehistoric to historic times.

Roman Funeral-urn, found at Watercrook

(*In Kendal Museum*)

Of the Roman roads we shall speak elsewhere, and at the same time refer to the more important of the Roman camps, which lie mainly along these roads. The chief Roman relics consist of pottery of very artistic types, some of which was made in England, though a large amount was imported from the continent, especially from Gaul. There are also many and various ornaments and a number

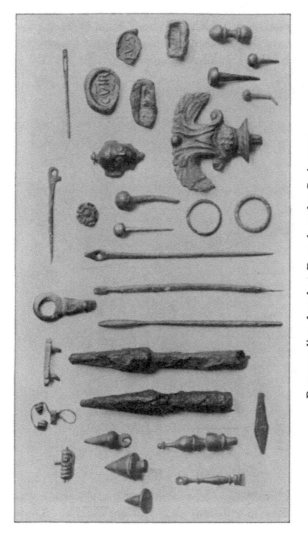

Roman relics, found at Brough-under-Stainmore

of Roman coins, which have been found in several places. We may mention also leaden seals, found in some abundance at Brough-under-Stainmore.

The Brough Stone

Inscribed stones have been found in places, for example a Roman milestone near Sedbergh and another near

Temple Sowerby: the most celebrated is the Brough Stone found at Brough-under-Stainmore, with a Greek inscription in memory of a Syrian youth, Hermes of Commagene.

In contrast with the number of Roman antiquities is the comparative paucity of relics of the Anglo-Saxons and the Norsemen. The relics of Norman and mediaeval times, apart from those connected with buildings, are also few.

17.   Architecture — (a)   Ecclesiastical. Churches and Religious Houses.

The ecclesiastical buildings of Westmorland are marked by a severe simplicity of style.   They are built of local materials: the more massive rocks of the slate group are chiefly used in the districts occupied by that group ; red sandstone in the Eden valley; and Carboniferous rocks, both sandstone and limestone, in the Carboniferous tracts. In the case of the more modern churches especially we frequently find Carboniferous rocks used for portions of the buildings even when the rest of the erection is of slate.

Westmorland contains no cathedral, for the whole of the county is in the diocese of Carlisle.

The larger churches are found at Appleby, Kendal, Kirkby Lonsdale, and Kirkby Stephen, and we find every gradation from these to the tiny churches which are so frequent in the county, such as those of Mardale Green, Haweswater, and Martindale, Ullswater.

The most characteristic feature of the churches is the low square tower, of which an admirable and simple example is that of Grasmere Church, seen in the illustration. The church usually consists of choir and nave, to

Grasmere Church

which in the case of the more important churches aisles have been added.

It is doubtful whether any remains of Saxon architecture are found in the churches of the county, though the Saxon churches of Tyneside form so noteworthy a feature in Northumberland.

Much work in many of the existing churches dates back to Norman times, with additions in the Tudor period.

Doorway, Kirkby Lonsdale Church

Many alterations have of course been made in the case of a large number of the churches at various later times, and a few churches are quite modern, as for example St Mary's, Ambleside. This possesses a spire, which is a marked

departure from the normal ecclesiastical architecture of
the county.

Of the religious houses the most important is Shap
Abbey, situate in the valley of the Lowther, over a mile
to the north-west of Shap. This abbey, which was a

Shap Abbey

house of an order of the Augustinian monks, was founded
in the twelfth century, and the church was begun soon
after the foundation, but the great western tower, which
is the chief surviving portion of the buildings, was erected
later, probably at the end of the fifteenth century. The
abbey was dissolved in the reign of Henry VIII.

Architecture—(*b*) **Military.   Castles.**

Though there is little impressive in connexion with the ecclesiastical architecture of the county, it is otherwise with its castles, many of which still exist as fine ruins. All were built by the Normans.   The greater number occur in the northern part of the county, to guard against

Brough Castle

invasion from Scotland.   These are said to have been originally erected by order of William Rufus, but much subsequent rebuilding with additions took place.   Two of them, at Brough-under-Stainmore and at Brougham, guarded the great road over Stainmore.   Another at Appleby protected that town, while a fourth known as Pendragon Castle, about four miles south of Kirkby

Stephen, guarded the Mallerstang pass at the head of the Eden valley.

The southern part of Westmorland was very efficiently protected from the raids of the northerners by the natural barrier of the fells through which the Shap pass forms a gap, and accordingly we find no military castles in this tract, for Kendal Castle is regarded as having been only the fortified dwelling-house of a Norman baron. It has been stated that it was built by Ivo de Taillebois in the reign of Stephen, but there is no definite evidence in favour of this statement.

The castles as first erected consisted of square towers or keeps as shewn in the accompanying illustrations of Brough and Brougham Castles, the keep of Brough being on the left of the picture, and that at Brougham being the high tower near the middle of the picture. Subsequently additions were made to the castles in the shape of a wall with subsidiary towers surrounding the ward or inner space in which stood the keep, while other subsidiary buildings might also be erected in this ward.

The castles of Brough and Brougham are built upon the old Roman earthworks, witnesses to the genius of the Romans for the selection of strategic positions, while those of Appleby and Pendragon are on earthworks of the pre-Norman thanes.

Brougham Castle is built on low ground by the side of the river Eamont. Appleby, Brough, and Pendragon Castles are on elevated ground; the first is on a loop of the river Eden, thus barring the end of the peninsula on which the houses of the town are built, and protecting also the houses of the bondsmen on the opposite side of the river. Brough is on a hill above Swindale Beck, and Pendragon stands on an eminence above the Eden.

Brougham Castle

Architecture — (c) Domestic. Manor Houses, Cottages.

The adoption of the Lake District as a resort for wealthy people from the adjoining counties has resulted in the erection of modern buildings of varied styles of architecture. With these we are not concerned, but shall consider only the more ancient buildings which present certain features typical of the district.

The more ancient dwellings may be divided into two

classes, the manor-houses occupied by the lords of the manors, and the cottages of the peasants.

The more interesting manor-houses were built between the fourteenth and the seventeenth centuries. The characteristic feature of these is the "peel-tower," rendered necessary for a people liable to border raids ; it originally constituted the whole dwelling. These towers

Yanwath Hall, with Peel-Tower

were modelled on the keeps of the Norman Castles. They were rectangular and usually three storied. In the lowest storey were kept provisions, the inhabitants occupied the middle storey by day and slept in the upper storey by night. The roof was used for fighting purposes when raiders from the north or from the Kent estuary had to be repelled. The peel tower stood in an enclosure called the "barmkyn," surrounded by a wall. Into this barmkyn the cattle were driven during the times of raids. In later times dining halls and other additions were built out from the peel tower.

Such towers cluster thickly around the Kent and its estuary. We find them for instance at Sizergh, Levens, Beetham, Dalham, and at Kentmere Hall. In northern

Levens Hall

Westmorland they occur scattered through the valley of the Eden and its tributaries. A fine group occurs in the neighbourhood of Penrith, including Clifton, Cliburn, and Yanwath. The last named, an admirable example of fifteenth century work, is shewn in the illustration with its attached buildings.

Levens is also famous for its garden, laid out by

James II's gardener, exhibiting one of the best examples of " topiary " work—the evergreens being trimmed into fantastic shapes.

The earlier cottages of the peasants were usually built of rough stone, often without mortar. They generally had three rooms on the ground floor, a sitting room and kitchen combined, a parlour, and a dairy. Stone steps,

Levens Gardens, shewing topiary work

which were often outside the house, led to a loft or sleeping-room. Various modifications naturally occur. Many of these cottages had round chimneys tapering towards the top, which form a characteristic feature of the old Westmorland cottage. The annexed illustration of an old cottage in Troutbeck near Windermere shews these chimneys.

Simplicity of construction characterises many of the bridges, especially over the smaller streams.

Old Cottage, Troutbeck

18. Communications—Past and Present.

In considering the lines of communication which have at different times been employed, two things must be taken into account. In the first place, the desire to proceed in a direct line from one point to another must, in the case of a hilly tract especially, undergo modifications in order to permit of easy travelling; hence the importance of the passes already insisted upon. Secondly there will be a tendency to follow the same lines at different periods of history, and accordingly many tracts and roads of different dates will occupy the same general lines, with minor modifications due to different modes of progression.

The pre-Roman occupants of the district, like the latter-day barbarians of other countries, no doubt had an intricate network of paths connecting village with village.

Bridge in Easedale

Such paths would be cut through the undergrowth of the lowlands, and in many cases extend over the higher ridges. Having no definite construction, those unused in later times would tend to disappear, while those which continued to be used, would now shew no signs of their use in pre-Roman days.

There is however some evidence of two important British routes in Westmorland, each of which was subsequently used. One of these came from the lowlands of the county of York over the Pass of Stainmore, and continued in a north-westerly direction down the Vale of Eden to Penrith and northwards into Cumberland. The other, from the south, went up the valley of the Lune, taking advantage of the natural gorge cut by that river through the high land of the Howgill Fells and the fells to the west of them ; it crossed the first road near Kirkby Thore, and went away over the Pennines into Northumberland.

It was the Romans who developed the great roads which opened up our county, and many of these are still in use, while even where they have fallen into disuse, their construction was such that they may often still be traced along considerable portions of their courses. The tracks of these Roman roads in Westmorland we must now consider. The more important of them are represented upon the map.

The most important Roman high-road in the county connected the Roman towns of Eboracum (York) and Luguvallium (Carlisle), and thus placed the fertile vale of Eden in communication with the east of England. Over Stainmore and down the vale of Eden to Kirkby Thore this road followed the line of the old British road before mentioned, utilising like it the depression through the Pennines. A little to the west of Lavatrae (Bowes) it leaves Yorkshire and enters Westmorland, a large camp, possibly of British origin but doubtless used by the

Romans, occurring close to the county boundary. At the summit of Stainmore a small four-sided Roman fort stands by the side of the road; it is known as the Maiden Castle. From this point the line of the road plunges

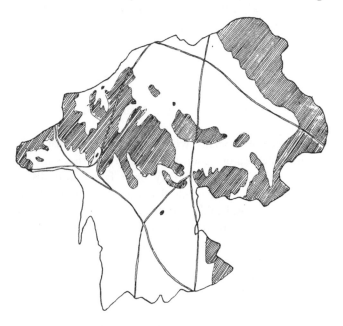

The Roman Roads of Westmorland
(Land over 1000 feet is shaded)

down into the vale of Eden to the Roman station of Verterae (Brough-under-Stainmore) where a large rectangular Roman camp exists. From Brough the road runs north-westward, past the camps of Crackenthorpe and Kirkby Thore (Brovonacae) to Brougham (Brovacum) on the Westmorland side of the Eamont, which here forms the county boundary. The highway thus formed the great connecting link between the fertile lowlands of north-east Westmorland and the civilised region to the south-east.

Another important road entered the county from the south near Kirkby Lonsdale, and connected the southern half of the county with the civilisation of southern England. By means of roads branching from this a great part of the county was opened up. Reaching the county, the road divided, one branch proceeding northwards, and the other to the north-west. The former connected north and south Westmorland by passing over the upland Shap pass, while the latter and its branches gave access to the hilly north-western parts of the county. We will first trace the course of the northern road, which follows the line of an old British road already mentioned.

From Kirkby Lonsdale it ran along the east side of the Lune to the neighbourhood of Sedbergh, where it crossed the river, and entering the gorge of the Lune proceeded to Low Borrow Bridge, where there is a Roman camp. Near Tebay it left the Lune gorge and ascended Shap Fell some way east of the present high-road over the pass, running down the north side of the fells to Kirkby Thore, where a camp stands by the road. At Kirkby Thore the first road is crossed, and access given by it to Brougham and Carlisle. North of this place the second road leaves the county near Kirkland and passes away over the fells of the Pennines to the great Roman Wall of Hadrian at Caervoran; going through the rich lead district of Alston Moor, which was probably worked by the Romans.

Returning to Kirkby Lonsdale and taking the route of the north-westerly road we are led to the Roman camp of Watercrook south of Kendal. Here the road is joined

The Gorge of the Lune, shewing the site of the Roman Camp at Borrow Bridge (marked by a x)

by one coming from the south, past Lancaster, and entering the county near Burton-in-Kendal, while a branch goes north-east to Borrow Bridge. From Kendal it goes to Ambleside, but near Staveley sends off a branch which joins another road from Ambleside to the north, to be noticed presently. A Roman station occurs near Ambleside, at the head of Lake Windermere. From this station

three other roads run.   One goes by Grasmere over the
pass of Dunmail Raise (where it leaves Westmorland) to
Keswick, and thence to the sea on the Cumbrian coast
to a point near the mouth of the Solway.   The second
road goes westward over the pass of Wrynose (where it
leaves the county) and that of Hardknott to the Roman
port of Ravonia (Ravenglass).   The third leaves Ambleside, extends in an easterly direction to Troutbeck, and
then passes northward, being soon joined by the branch
from near Staveley already noticed.   It mounts the hills
near the head of Troutbeck, and continues on or near
the top of the ridge of the High Street range (which
obtains its name of High Street from this circumstance)
until the ridge sinks down into the plain near the foot
of Ullswater, after which the road continues on lower
ground to Brougham.   The illustration on the next page
shews part of the course of this road.   At this point
the road is over two thousand feet high, and it lies
above this level for many miles, still traceable, owing
to the excellence of its construction, over considerable
portions of its course along the ridge.   Every one who
can, should ascend High Street from Troutbeck and walk
along this upland road to get some idea of the influence
of the Romans upon this northern tract so far removed
from the central seat of civilisation at Rome.

  After the wane of Roman influence, no important
changes in the means of communication occurred until
the introduction of railways.   The main Roman routes
still continued to be trodden, though in some cases slight
deviations from the former roads were made.   Through

the middle ages men still came over from the east by the
Stainmore route into the Eden valley. The entrance
from the south was either along the old road from Kirkby
Lonsdale up the Lune valley, or by the route from
Lancaster to Watercrook : thence either to Tebay on
the north-east; Ullswater on the north by a road over
Kirkstone Pass parallel with the old High Street road

The Roman Road on High Street (marked by crosses)

(though there is reason to believe that a minor Roman
road ran over Kirkstone); to Ambleside, and through
Dunmail Raise to Keswick on the north-west; or to
Ravenglass on the west by Wrynose and Hardknott
passes. In each of these cases the passes frequented in
the early days were still utilised. Some changes were no
doubt made. For instance, after passing over the Shap
Pass, the direct road northward to Penrith goes through

Shap, and not round by Kirkby Thore, though even here the probability is that a road of less importance than those which we have described existed in Roman times.

A network of new routes sprang up as the area became more populous. They probably began as foot-paths and bridle-tracks, and some of them by degrees became used by vehicles. Even in the case of the latter

The Kendal and Lancaster Canal

their difference from the old Roman roads is usually easily detected, for they are generally badly graded, and it is only here and there that they depart from the old tracks.

The old pack-roads were used for carrying produce from the farms and hamlets to the nearest market towns. Such as have not been converted into modern roads have in many cases disappeared. As the valley bottoms were often thickly covered with undergrowth which could with difficulty be penetrated, the pack-roads were frequently taken up the sides of the fells from the houses in the valleys, and accordingly we still meet with relics of these pack-roads on the fells where there is now little or no traffic, and they usually take advantage of the smaller passes, which indent the ridges between valley and valley.

In 1819 a canal was opened between Kendal and Lancaster, which was in communication with the canal system existing further south. At one time a passenger packet-boat plied on this in addition to the barges used for carrying goods. This canal is still in use.

We now turn to the railways. Here, again, advantage has been taken of the great passes which were used by the Romans for their chief roads, and the railways are carried through them. The London and North Western, Midland, North Eastern, and Furness Railway Companies have lines within the county. The earliest railway in the county, the Lancaster and Carlisle line—afterwards taken over by the London and North Western Company—was made in 1844–46.

The North Western main line enters the county from the south near Burton-in-Kendal, and runs parallel with the old Roman road to Watercrook until it reaches Oxenholme; thence it is continued not far from another Roman road to Lowgill Junction where it joins the course of the great road from Kirkby Lonsdale over Shap Fells. From Shap Fells it runs parallel with the coach-road to Penrith. It has two branch lines; one again follows the Roman road from Oxenholme to Kendal and Windermere, the other, from Lowgill to Ingleton, goes down the Lune valley, following again the Kirkby Lonsdale road but in the opposite direction to that of the main line.

The Midland main line enters the county in the south near the source of the Eden, taking advantage of the passes between the Eden and Ure, and between the latter stream and the Clough. It is carried down the

Steamer on Windermere

upper Eden valley (Mallerstang) to Kirkby Stephen, and from thence continues in the expanded Vale of Eden to the point where it leaves the county near Newbiggin, north of Appleby.

The North Eastern line joins the North Western at Tebay. It runs up the Lune to Ravenstonedale, and then takes advantage of a curious gorge where the former head-waters of the Lune have naturally been diverted into the Eden between Ravenstonedale and Smardale. At Smardale Station it issues from this gorge into the main Eden vale, which it crosses at Kirkby Stephen, and then climbs the Pennine hills over Stainmore, rising to a height of 1369 feet.

From Kirkby Stephen a branch line of this system goes down the Eden valley to Appleby, and leaving the

The Ferry-boat, Windermere

valley at Kirkby Thore is carried over low ground to Penrith, where the North Eastern makes another junction with the North Western system.

The main Furness line enters Westmorland near Arnside, and leaves the county about four miles to the west of this before reaching Grange. In this short distance it crosses over the sands of the Kent estuary by a fine viaduct. A branch line from Arnside is carried to the north-east to join the North Western line at Hincaster, south of Oxenholme. The Furness line also runs a service of steamers up Windermere, and a steamer plies on Ulls-water connected by coach with the London and North Western Railway at Penrith.

In the part of the district frequented by tourists, coaches are still much employed for their conveyance, but motor vehicles are now used for the public conveyance of passengers on some of the routes, as for instance on the Windermere-Keswick Road, and they will no doubt become much more frequent.

With all these changes in the method of travel, it is interesting to find that the principal routes are in the main those of the old Roman roads, which were skilfully planned along the lines of least resistance when the physical character of the country was taken into account.

19. Administration and Divisions.

We have seen that Westmorland was definitely formed into a county by Henry I, but that in pre-Norman times the Saxons had divided southern tracts into shires. These shires were "shorn" off from larger tracts of country for administrative purposes.

In the early days of tribes administration was largely a family affair, and later an affair of clans, but as the tracts of land under one ruler increased it was necessary to have definite divisions, in each of which administration of some of its affairs was local, and the administration according to families and clans was replaced by a territorial one. The early Saxon "shires" of the south, each probably the result of expansion outwards of a definite colony, formed convenient territorial areas for administrative purposes, and under the Normans those areas which had not thus been parcelled out were formed into counties. Hence the existence of the county of Westmorland for similar purposes.

The Saxons had *ealdremen* or governors, who appointed deputies called *sheriffs* (shire-reeves ; reeve being equivalent to our word steward). The inferior people were partly *ceorls* or freemen, and partly *villeins*, who were labourers in the service of particular persons, and not strictly slaves. Upon the establishment of the feudal system by the Normans many of the Saxon laws and customs were retained, as was also the old distinction of classes. Thus there were counts or earls, barons, knights, esquires, free tenants, and villeins. When Henry I formed the separate counties of Cumberland and Westmorland the administrative power was put into the hands of Sheriffs.

The county of Westmorland upon its establishment consisted of two baronies, that of Appleby (or, as it was at first called, the Barony of Westmorland), and that of Kendal. The division between these baronies roughly coincided with the high land of the Shap Fells, the barony of Appleby being in the north-eastern and that of Kendal in the south-western part of the county.

Those responsible for the administration of the affairs of the county saw to the collection of taxes for the Crown, the supply of soldiers for military service, and the administration of justice within the county. It will thus be seen that the county is an area which was separated for the purposes of internal government, but it also took some part in the affairs of the nation.

The system of local government gave rise to still smaller divisions for administrative purposes. Thus each of the baronies was divided into *hundreds* or *wards*; that of Kendal contained the Kendal and Lonsdale Wards, and that of Appleby the East and West Wards.

A further division was made into *parishes*, each with its own officials, and the parishes were again sub-divided into *townships* or *constablewicks*. As the shire had its sheriff so the parish had its own special reeve, or presiding official.

The gradual accumulation of people in special areas giving rise to towns necessitated special government in the case of these towns apart from the constable of an ordinary constablewick with his subordinates. The first charter was granted to Appleby in 1179, wherein certain grants were made to the burgesses of the town. There was at that time no mayor and it is probable that the first officer of the town under the name of town *reeve*, *provost*, or some such title, had a jurisdiction over the town which would be somewhat akin to that of the sheriff over the county. The exact date when Appleby possessed a mayor is unknown, but it was some time in the early part of the thirteenth century.

Kendal did not become a corporate town until 1575, when a charter was granted and the town governed by an alderman and burgesses. In the second year of Charles I

the town obtained a fresh charter, and therein the new
governing body was headed by a mayor.

When the county was first constituted its voice in the
general affairs of the nation was slight.   When the great
Charter was signed in 1215 one of its provisions was that
its articles were to be carried out by twelve sworn knights
from each shire chosen in the County Court.   Thus the
influence of the county in national affairs became more
direct.   In 1295 the first complete Parliament assembled,
and, besides others, two knights were summoned from
each shire, two citizens from each city, and two burgesses
from each borough.   Since then Westmorland has had its
full share in the government of the nation, and after many
changes it is now represented by two members for tracts
of the county which are once more largely determined by
the natural barrier of the Shap Fells, for one member sits
for the Appleby division, the other for the Kendal division
of the county.

Let us turn now to the present government of the
county, which has gradually grown out of the old administrative system.   The head officer of the county is the
Lord Lieutenant, who in some ways represents his Norman
predecessor who, under the title of count, earl, or other
name, was at the head of affairs.   The Lord Lieutenant
represents the Crown in the county and one of his duties
is to nominate all Deputy Lieutenants and Justices of the
Peace.

The High Sheriff of to-day is to some extent repre-
sentative of the Norman Sheriff, although his duties are
much restricted in comparison with those of the ancient

officer, and are largely connected with affairs of the law. He is Keeper of the King's Peace within the county, and he attends the Judges of the Realm when on Circuit.

A recently constituted body with purely administrative functions is the County Council. It consists of a Chairman, Vice-Chairman, fourteen Aldermen and forty-two elected Councillors. Among its functions are the management of county halls and buildings, pauper lunatic asylums, bridges and main roads, the appointment of certain officers such as coroners, the control of parliamentary polling districts, and contagious diseases of animals. The County Council are also the local education authorities through an Education Committee. There are also Rural District Councils and Urban Councils as well as Parish Councils, for the administration of smaller areas of the county.

For purposes of Justice the county has Assizes which are held at Appleby, and Quarter Sessions which are held at Kendal, and also a number of Petty Sessions, each having Justices of the Peace, whose duty it is to try and to punish offenders against the law.

For Ecclesiastical purposes the county is in the Diocese of Carlisle of the Province of York.

20.   The Roll of Honour of the County.

Lying, as it does, far from the great centres of commerce or learning, both in earlier and modern times, and with a small and scattered population living in mountainous districts with no facilities of intercommunication, and the highly educated classes in great minority, Westmorland could hardly be expected to shew the long list of famous men that Kent, Norfolk, and other such favoured counties produced. At the same time she

William Wordsworth

can claim a good number who have made themselves famous and honoured names.

Foremost among those born in the county we may place Catherine Parr, sixth and last wife of King Henry VIII. She was the daughter of Sir Thomas Parr, and is stated to have been born in Kendal Castle in 1513. This Queen, who died in 1548, wrote some religious works.

A large number of distinguished ecclesiastics claim Westmorland as the county of their birth. This is no doubt in part due to the scholarships at Queen's College,

Dove Cottage, Grasmere
Occupied by Wordsworth and, later, by De Quincey

Oxford, which are open to those who were born in the county. Two of these ecclesiastics attained to a very high dignity, namely Christopher Baynbrigg and Hugh Curwen. The former was apparently born at Hilton, near Appleby, in 1460. He became Archbishop of York and afterwards a cardinal. He died in 1514. Curwen was born in 1500. His birthplace is unknown, but there

is evidence that he was a native of Westmorland. He
became Archbishop of Dublin and Lord Chancellor of
Ireland, and died in 1567.

The county can boast of many bishops, but greater
than any was a man who refused the Bishopric of Carlisle
to which he was nominated by Queen Elizabeth. This

Rydal Mount, Wordsworth's later home

was Bernard Gilpin, known as "the Father of the Poor,"
and more generally as "the Apostle of the North." He
was born at Kentmere Hall in 1517, and devoted his life
to promoting the great work of the Reformation in the
north of England until his death in 1583.

Among the contributors to literature we may notice

John Mill (born 1645, died 1707), of Hardendale, near Shap, who became Principal of St Edmund's Hall, Oxford, and brought out an edition of the New Testament in Greek. John Langhorne (1735—1779), Prebendary of

De Quincey

Wells, of Winton, near Kirkby Stephen, a minor poet. Richard Burn (1709—1785), born also at Winton, who was Chancellor of Carlisle, was a well-known antiquary, and with Joseph Nicholson wrote *The History and Antiquities of the Counties of Cumberland and Westmorland.*

John Hodgson (1780—1845), born in Swindale near Shap, was the historian of Northumberland. Ephraim Chambers (1680—1740), of Heversham, was the compiler of the well-known *Chambers' Encyclopaedia.* Thomas Shaw (1692—1751), born in Kendal, became Principal of St Edmund's Hall, Oxford, and was a celebrated traveller in northern Africa and the Holy Land.

Of men of science we may notice Thomas Garnett (1766—1802), who was born at Casterton and became first Professor at the Royal Institution, and John Gough (1757—1825), born in Kendal, a blind mathematician and natural philosopher. Kendal may be justly proud of the number of devotees of science whom she owns. Among them were two botanists—John Wilson (1702—1751), author of a *Synopsis of British Plants,* and William Hudson (1734—1793), author of the *Flora Anglica.*

Sir John Wilson (1741—1793), of Applethwaite, and Sir Alan Chambre (1740—1823), of Kendal, became judges; Sir Thomas Bowser (1748—1833), of Kirkby Thore, was Commander-in-chief of the Madras Army and wrote *Memoirs of the Late War in Asia* in 1788. Sir Richard Pearson (1731—1805), born near Appleby, was an Admiral in the Navy, and is celebrated for his defence of the Baltic Fleet against the American squadron under Paul Jones in 1779.

Though born in the adjacent county of Cumberland, one name may well be included in the Roll of Honour of Westmorland, that of William Wordsworth, the poet (1770—1850), for a great part of his life was spent in the vale of the Rothay, and it was here that most of his work was done. Great as a poet, he must also be regarded as an important factor in the prosperity of the county, for the recognition of the Lake District as a place of beauty owes much to his writings.

During Wordsworth's lifetime and afterwards, the Vale of Rothay was an intellectual centre. Many

Fox How, Dr Arnold's house near Ambleside

eminent men, as Coleridge, Sir Walter Scott, and Southey paid fleeting visits to Wordsworth, but others made their abode here for longer time. We may mention two men of very different natures, De Quincey the writer, and Dr Arnold, the celebrated Rugby Headmaster.

21. THE CHIEF TOWNS AND VILLAGES OF WESTMORLAND.

(The figures in brackets give the population in 1901, the asterisk denoting parishes. The figures at the end of each section are references to the pages in the text.)

Ambleside (2536*), a market-town in the Kendal ward, 25 miles west-south-west of Appleby and a mile from the head of Windermere, contained formerly some manufactories of linsey-woolsey. The main part of the town is modern, and has arisen on account of the importance of the place as a tourist-centre, for visitors arrive at Windermere station by train and thence proceed by road to Ambleside, or they come by the Furness railway to the lake foot and thence by the company's steamers to Ambleside. Coaches connect the town with the Coniston, Langdale, Grasmere, Keswick and Ullswater regions. The ancient church is just outside the town, the modern church with its spire was built according to the plans of Sir Gilbert Scott. The market day is Wednesday. (pp. 28, 62, 82, 90, 93, 111, 124—126.)

Appleby (1764), the county-town of Westmorland, is mostly built upon a small peninsula nearly surrounded by the river Eden. Bongate, however (the former dwelling-place of the old bondsmen), is on the east bank of the river and since the extension of the Midland railway the town has undergone further development on this side. The main street extends up the hill, having the castle at its upper and the church of St Lawrence at its lower end. In front of the church is the market-square, with the old Moot Hall in the place where the burghers used to assemble for their "mote" or meeting for the transaction of public business. Near it is still the old bull-ring used for the cruel sport of bull-baiting. At the upper end of the street is the Hospital of St Anne, founded by Anne, Countess of Pembroke, for thirteen widows of tenants. Formerly a house of the Carmelite monks or White Friars

Waterhead, Ambleside

existed, but no trace of it now remains. In the thirteenth and the early part of the fourteenth centuries the town appears to have been of great importance, but it never recovered from the attack made by the Scots in 1388. The town has a mayor and corporation. There is a weekly market on Saturdays, and two or three fairs in the course of the year. The Assizes are held here. (pp. 2, 34, 42, 44, 63, 81, 99, 109, 113, 114, 129, 130, 132—135.)

Arnside, a small but rising watering-place on the left bank of the Kent estuary. An interesting phenomenon here is the *bore* or tidal wave coming up the estuary.   (pp. 8, 36, 38, 46, 63, 66, 130.)

Brough-under-Stainmore (921*), a town in the east ward of Westmorland, eight miles south-east-by-east from Appleby, is situated on Swindale Beck, a tributary of the river Eden, in a rich agricultural district.   The castle is a fine Norman keep described on p. 114.   There are several fairs in the course of the

Appleby and the Pennines

year, including the celebrated Brough Hill Fair, held on the last day of September and the first of October.   (pp. 35, 107—109, 113—115, 122.)

Burton-in-Kendal (703*), a market-town in the Lonsdale ward, consists chiefly of one long street and the market square. It is 34¼ miles south-west-by-south from Appleby.   The church is dedicated to St James.   Markets are held on Tuesday.   (pp. 10, 30, 38, 60, 124, 128.)

Kendal (14,183) is situated on the river Kent, the greater part of the town being on the right (west) bank on ground rising up to the limestone eminence of Kendal Fell. It is by far the largest town in the county. It lies 23 miles south-west-by-west from Appleby. The northern portion is known as Stricklandgate and the southern as Highgate, and where these unite a third street extends north-eastwards to the Kent—the Stramongate, from which Stramongate Bridge leads to that part of the town which lies east

Kendal

of the river. A series of terraced streets rise high up the fell-side. The fine parish church dedicated to the Holy Trinity consists of nave, chancel, and fine side aisles, and has a low square tower. At the northern end is St Thomas' church. The Grammar School was founded in 1525. The town contains an excellent Museum illustrating the natural history and antiquities of the district. The ruins of the castle are on a hill on the east side of the river. The town has a mayor and corporation. The

Quarter Sessions are held in Kendal, and there is a market every Wednesday and Saturday. (pp. 26, 30, 35, 38, 46, 60, 89, 93, 94, 101, 106, 109, 114, 124, 127, 128, 132—135, 137, 140.)

Kirkby Lonsdale (1871*), 30 miles south-by-west from Appleby, stands on high ground on the west bank of the Lune. It is on a branch line of the London and North Western Railway. The interesting church of St Mary has fine aisles and contains much Norman work. The Grammar School was founded

Kirkby Lonsdale

in 1591. (pp. 26, 30, 35, 38, 45, 94, 95, 105, 109, 123, 126, 128, 129.)

Kirkby Stephen (1656), 11 miles south-east-by-south from Appleby, is on the left bank of the Eden, with two railway stations, one on the North Eastern and the other on the Midland Railway. St Stephen's is a fine church containing interesting monuments of the Wharton and Musgrave families. The

Old Cross at Kirkby Lonsdale

Grammar School was founded in 1558. Markets are held every Monday. (pp. 26, 30, 34, 42, 43, 93, 109, 113, 129, 130.)

Milnthorpe (1051*), 32 miles south-west-by-south from Appleby, is on the northern bank of the river Bela where it enters the estuary of the Kent and is the only port of Westmorland, though now of no importance as such. The parish church is at Heversham, which adjoins Milnthorpe. At Heversham also is the largest Grammar School in the county, founded 1613. Milnthorpe Station is on the main London and North Western line, and Heversham Station on a branch of the Furness line. (pp. 5, 12.)

Orton (832*), nine miles south-west-by-south from Appleby, is situated on the southern slope of Orton Scar at an elevation of about 800 feet above sea-level. The church is plain with a solid square tower.

Shap (1490*) is a village of some size, consisting chiefly of one long street along the north road between Kendal and Penrith. The church is dedicated to St Michael. The abbey has already been noticed, and mention made of the numerous prehistoric antiquities of the neighbourhood. A medicinal spring rises about four miles south of the village. (pp. 23, 26, 41, 44, 93, 94, 105, 112, 123, 126.)

Windermere Village with **Bowness-on-Windermere** are of importance as a tourist centre. They are practically united to form one long village with numerous hotels and lodging-houses. Bowness is built on Bowness Bay and the rising ground around, and has a station for the steamers which ply on the lake. The church stands close to the shores of the lake. (pp. 90, 105, 118, 128, 131.)

Fig. 1. Population of Westmorland (64,409) compared
with that of England and Wales

Fig. 2. Area of Westmorland (505,330 acres) compared
with that of England and Wales

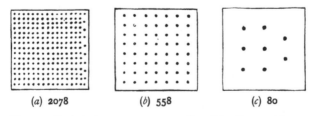

Fig. 3. Population per square mile of (a) Lancashire,
(b) England and Wales, and (c) Westmorland

(*Each dot represents 10 persons*)

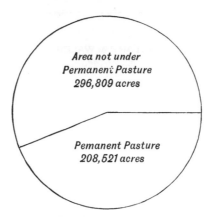

Fig. 4. Area of permanent pasture compared with rest of Westmorland

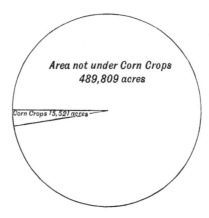

Fig. 5. Area under cereals compared with rest of Westmorland

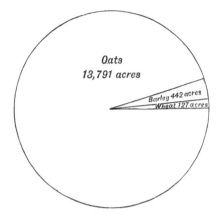

Fig. 6. Comparative areas of cultivation of oats, barley, and wheat in Westmorland

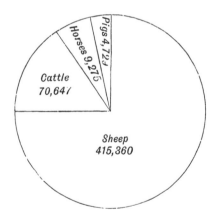

Fig. 7. Comparative numbers of sheep, cattle, horses, and pigs in Westmorland